PESTICIDE-RELATED ILLNESS AND INJURY SURVEILLANCE

A How-To Guide for State-Based Programs

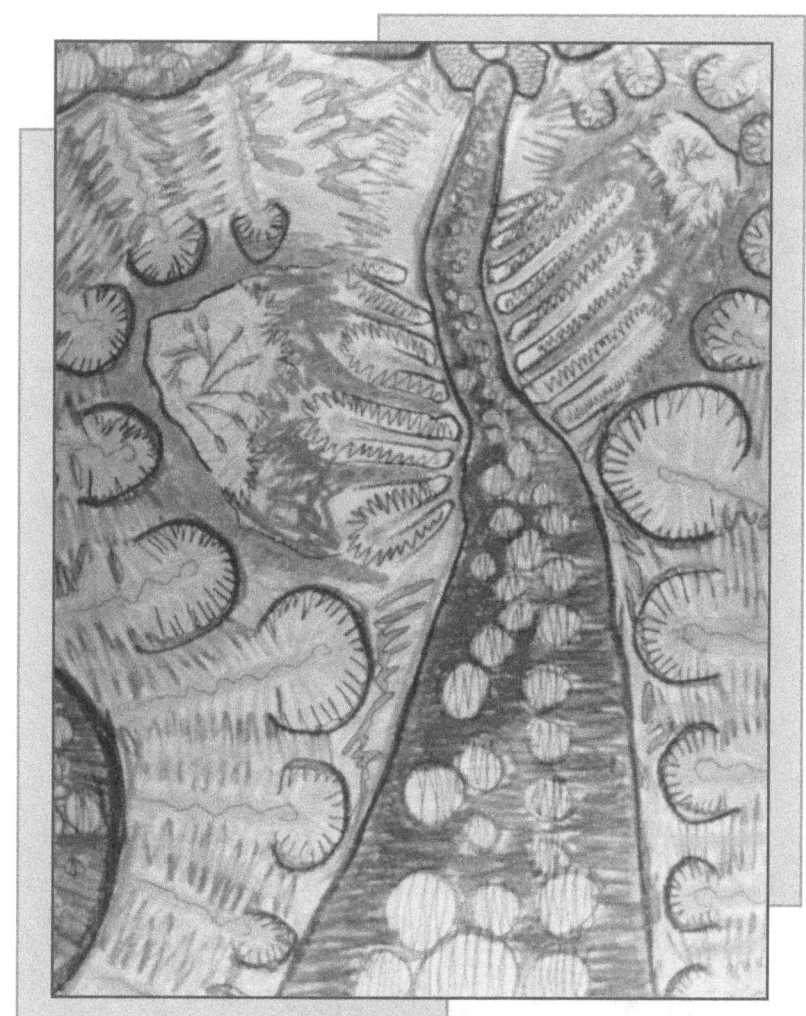

DEPARTMENT OF HEALTH AND HUMAN SERVICES
Centers for Disease Control and Prevention
National Institute for Occupational Safety and Health

DISCLAIMER ■

Mention of any company or product does not constitute endorsement by the National Institute for Occupational Safety and Health (NIOSH). In addition, citations to Web sites do not constitute NIOSH endorsement of the sponsoring organizations or their programs or products. Furthermore, NIOSH is not responsible for the content of these Web sites.

ORDERING INFORMATION

To receive documents or other information about occupational safety and health topics, contact NIOSH at

NIOSH
Publications Dissemination
4676 Columbia Parkway
Cincinnati, OH 45226–1998

Telephone: **1–800–35–NIOSH** (1–800–356–4674)
Fax: 513–533–8573
E-mail: pubstaft@cdc.gov

or visit the NIOSH Web site at **www.cdc.gov/niosh**

This document is in the public domain and may be freely copied or reprinted.

DHHS (NIOSH) Publication Number 2006–102

October 2005

Foreword

Surveillance data can be used to identify new emerging pesticide problems, estimate the magnitude of pesticide poisoning, and evaluate intervention and prevention efforts. Recognizing this, the National Institute for Occupational Safety and Health (NIOSH) Strategic Surveillance Plan recommends that States conduct surveillance for acute pesticide-related illness and injury.

Since 1987, NIOSH has provided financial and technical support for State-based acute pesticide poisoning surveillance programs. NIOSH is not the only organization that has recommended improved and/or expanded surveillance in this area. Others include the American Medical Association, the Council for State and Territorial Epidemiologists, the United States Government Accountability Office, and the Pew Environmental Health Commission. Despite these recommendations, most States do not conduct acute pesticide-related illness and injury surveillance.

Acute pesticide-related illness is a relatively complex disease. Approximately 16,000 pesticide products are currently registered in the United States. In addition, all organ systems are susceptible to pesticide toxicity. The multitude of pesticide products and associated health effects may act as a barrier to establishing surveillance programs. NIOSH developed this guide to provide standards and principles that can help to master this complexity.

We expect this document will be useful to agencies that are developing an acute pesticide-related illness and injury surveillance program or are interested in maintaining and improving an established surveillance program. The guide provides (1) information about the importance of pesticide poisoning surveillance; (2) mechanisms to improve reporting of cases to surveillance programs; (3) methods to investigate reported cases; (4) guidance on using the case definition; and (5) additional resources on pesticide toxicology, pesticide usage, governmental partners, and surveillance.

To our knowledge, this is the most comprehensive instruction guide for pesticide-related illness and injury surveillance. The goal of this guide is to assist the efforts of our partners to identify pesticide poisoning risk factors. Pesticide poisoning prevention can be achieved by targeting interventions toward these identified risk factors. NIOSH hopes individuals and agencies interested in pesticide poisoning surveillance and prevention (e.g., local, State, and Federal government agencies, community-based organizations, and international agencies) will find this guide useful for identifying and preventing pesticide poisoning.

John Howard, M.D.
Director
National Institute for Occupational
 Safety and Health
Centers for Disease Control
 and Prevention

Contents

Ordering Information ii
Disclaimer ii
Foreword iii
Abbreviations xi
Acknowledgments xiv
1. Introduction 3
   1.1   A Guide to the Manual 3
          1.1.1   Goal 3
          1.1.2   How to Use the Manual 3
          1.1.3   Limits of the Manual 3
   1.2   Introduction to the Problem 4
          1.2.1   Problem Overview 4
                  1.2.1.1   The Changing Patterns of Pesticide Use 4
                  1.2.1.2   Scope of the Problem 4
                  1.2.1.3   Susceptibility to Pesticide Poisoning 5
                  1.2.1.4   A Problem That Affects Us All 5
          1.2.2   Why Investigate Reports of Pesticide Poisoning? 5
          1.2.3   Response to the Problem 6
2. First Steps in Surveillance System Design: Objectives, Resources, and the Reporting Rule . . 11
   2.1   Introduction 11
   2.2   Objectives of Pesticide Poisoning Surveillance 11
   2.3   Program Staffing and Structure 12
          2.3.1   Types of Expertise Needed for Surveillance 12
          2.3.2   Ways to Meet Needed Expertise with Minimal Resources and Staffing Levels . 12
   2.4   Program Funding Options 13
   2.5   Reporting Requirements and Rules 13
          2.5.1   Elements of the Reporting Rule 15
                  2.5.1.1   What Is Reportable? 15
                  2.5.1.2   Who Must Report? 16
                  2.5.1.3   Where to Report 17
                  2.5.1.4   When to Report 17
                  2.5.1.5   Health Insurance Portability and Accountability Rule and Public
                                  Health (HIPAA) Privacy Rule and Public Health Surveillance 18
                  2.5.1.6   Confidentiality 19
                  2.5.1.7   Interagency Cooperation or Sharing of Information 19
                  2.5.1.8   Authority to Investigate 19
          2.5.2   Examples of Reporting Laws and Rules 20
                  2.5.2.1   New York Reporting Rule 20
                  2.5.2.2   Washington Law 20
                  2.5.2.3   Texas Reporting Law 20

2.6   Surveillance Strategy for States with Limited Resources 20
3. Case Ascertainment 27
    3.1   Introduction 27
    3.2   Poison Control Centers (PCCs) 27
    3.3   Workers' Compensation Data 29
    3.4   Health Care Professionals (HCPs) 29
    3.5   Referral from Other Agencies 32
    3.6   Emergency Department Logs 32
    3.7   Affected Persons 32
    3.8   Worker Representatives 33
    3.9   Hospital Discharge Data (HDD) 33
    3.10 Laboratories 33
4. Data Collection and Management 37
    4.1   Introduction 37
    4.2   Data Standardization 37
    4.3   Documentation of Procedures 37
    4.4   Data Collection 38
            4.4.1   Standard Variables to Be Collected by Pesticide Poisoning Surveillance
                    Program (PPSPs) 38
                    4.4.1.1 Administrative and Demographic Variables 38
                    4.4.1.2 Occupation and Industry Data 39
                    4.4.1.3 Exposure Descriptions 39
                    4.4.1.4 Chemical Information 40
                    4.4.1.5 Health Effects Descriptors 40
                    4.4.1.6 Investigation Findings 40
                    4.4.1.7 Case Classification 40
            4.4.2   Optional Variables 40
            4.4.3   Introduction to the SPIDER Program 41
    4.5   Data Management 42
            4.5.1 General Guidelines for Data Management 42
                    4.5.1.1 Confidentiality and Security 42
                    4.5.1.2 System Backups 42
                    4.5.1.3 Transmitting Data to NIOSH 43
5. Case Intake and Follow-up 47
    5.1   Introduction 47
    5.2   Overview of the Case Investigation Process 47
    5.3   Initial Report Intake (Complaint Evaluation) 48
    5.4   Case Follow-up Interviews 50
            5.4.1   Affected Persons 51
                    5.4.1.1 Chemically Sensitive Persons 53
            5.4.2   HCPs 53

		5.4.3 Third Parties-Pesticide Applicators, Landlords, and Employers 54
5.5	Notification of the Local Health Department 55
5.6	Obtaining Pesticide Product Information 55
	5.6.1 Inert Ingredients 56
5.7	Evaluation and Referral for Site Inspection or Enforcement 57
5.8	Overview of Agencies with Jurisdiction of Pesticides 58
	5.8.1 Federal Insecticide, Fungicide, and Rodenticide Act (FIFRA) State Designees 58
	5.8.2 Occupational Safety and Health Administration (OSHA) 59
	5.8.3 Agencies Responsible for Disease Surveillance and Control 59
	5.8.4 Other State and Adjunct Agencies 59
		5.8.4.1 Vector Control Districts 60
		5.8.4.2 Pest Control Boards 60
5.9	State Interagency Coordination of Case Investigations 60
	5.9.1 Interagency Agreements 60
	5.9.2 Multiagency Coordinating Boards 60
	5.9.3 Advisory Committees 61
5.10	NIOSH, National Center for Environmental Health (NCEH), and the U.S. Environmental Protection Agency (EPA) 61
5.11	Federal Agencies That May Have a Role or Be a Resource During Case Investigation 62
	5.11.1 United States Department of Agriculture (USDA) 62
		5.11.1.1 Cooperative State Research Education and Extension System (CSREES) 62
		5.11.1.2 Federal Grain Inspection Service (FGIS) 62
		5.11.1.3 Animal Plant Health Inspection Service (APHIS) 63
	5.11.2 Federal Aviation Administration (FAA), National Transportation Safety Board 63
	5.11.3 U.S. Fish and Wildlife 63
	5.11.4 Other Federal Agencies 64
		5.11.4.1 Coast Guard 64
		5.11.4.2 Consumer Product Safety Commission (CPSC) 64
		5.11.4.3 Customs Bureau 64
		5.11.4.4 Department of Transportation (DOT) 64
		5.11.4.5 Federal Bureau of Investigation (FBI) 64
		5.11.4.6 Federal Railway Administration 64
		5.11.4.7 Food and Drug Administration (FDA) 64

6. Site Inspections by PPSP 67
   6.1 Introduction 67
   6.2 Getting Started with the Inspection 68
   6.3 Site Walk-Around Evaluation 69
   6.4 Equipment for Site Inspections 69
       6.4.1 Camera 69

		6.4.2	Personal Protective Equipment (PPE) 69
		6.4.3	Water 70
		6.4.4	Contact Form 70
	6.5	Sampling 70
		6.5.1	Sample Collection 71
		6.5.2	Sample Preparation Custody and Handling 71

7. Case Closure and Classification 75
7.1 The Case Closure Process 75
7.2 Case Definition for Acute Pesticide-Related Illness and Injury Cases 75
7.3 Clinical Description 75
7.3.1 Case Definition 75
7.3.2 Laboratory Criteria for Diagnosis 76
7.3.3 Case Classification Categories 77
7.4 Communication of Findings and Recommendations 77

8. Data Analysis and Reporting of Aggregated Data 81
8.1 Introduction 81
8.2 Routine Descriptive Analysis 81
8.2.1 Person-Based Analyses 81
8.2.1.1 Case Series 81
8.2.1.2 Bivariate Analyses 81
8.2.1.3 Rates 82
8.2.1.4 Time-Based Analyses 83
8.2.1.5 Place-Based Analyses: Mapping of Data 83
8.2.2 Data Dissemination: Defining What Reports Will Be Useful For a Specific Program 84

9. Developing Intervention Strategies and Evaluating Surveillance 87
9.1 Designing Interventions Based on Surveillance Data 87
9.1.1 Education 87
9.1.1.1 At Home and at Work 87
9.1.1.2 Educating the Educators 88
9.1.1.3 Educating the General Public 88
9.1.2 Engineering Controls, Modifications 88
9.1.2.1 Closed Mixing and Loading Systems 89
9.1.2.2 Creative Formulations and Technologies 89
9.1.2.3 Enclosed Cabs and Other Protection 89
9.1.2.4 Application and Equipment Modifications 89
9.1.3 Administrative Controls and Regulatory Changes 89
9.1.4 Examples of Interventions 90
9.1.4.1 Automatic Insecticide Dispensers 90
9.1.4.2 Mevinphos 90
9.1.4.3 Bis(tributyltin)oxide (TBTO) Paint Additives 91

9.2	Evaluation of the Surveillance System	92
9.3	Conclusion	93

References 97

Appendix A. Listing of Morbidity and Mortality Weekly Report (MMWR) Articles on Pesticide-related Illness and Injury; January 1, 1982–September 30, 2004 105

Appendix B. Selected Pesticide Illness Reporting Statutes and Rules 111

Appendix C. Sample Forms, Investigation Tools, and Templates for Data Tables 115
    C.1   Case Tracking Form and Contact Log 116
    C.2   Main Pesticide Exposure Questionnaire 119
    C.3   Pesticide Illness and Injury Surveillance Data Collection Form 141
    C.4   Field Investigation Contact Form and Health Safety Checklist for Field Personnel ..161
    C.5   Instructions for National Transportation Safety Board (NTSB) Search to Obtain Reports of Airplane Accidents Involving Aerial Pesticide Applicators 165
    C.6   Sample Templates for Tables Presenting Surveillance Data 166
    C.7   Sample Letters for Case Follow-up 169
    C.8   Instructions for Obtaining Acute Pesticide-related Illness and Injury Reports from Poison Control Centers (PCCs) 176

Appendix D. Case Definition for Acute Pesticide-Related Illness and Injury Cases Reportable to the National Public Health Surveillance System 183
    D.1   Clinical Description 183
    D.2   Laboratory Criteria for Diagnosis 184
    D.3   Classification Criteria 184
    D.4   Contacts for Additional Information 187
    D.5   Frequently Asked Questions (FAQs) 187
    D.6.   Characteristic Signs and Symptoms for Several Pesticide Active Ingredients and Classes of Pesticides 192
    D.7   Glossary of Medical Terms 204

Appendix E. Severity Index for Use in State-based Surveillance of Acute Pesticide-related Illness and Injury 209

Appendix F. Pesticide Law and Definitions 217
    F.1   Federal Insecticide, Fungicide, and Rodenticide Act (FIFRA) 217
    F.2   Federal Food, Drug, and Cosmetic Act (FFDCA) 223
    F.3   Food Quality Protection Act (FQPA) 223
    F.4   Safe Drinking Water Act (SDWA) 224
    F.5   Occupational Safety and Health Act (OSH Act) 224

Appendix G. Resources for Additional Information Related to Pesticide Poisoning Surveillance 229
    G.1   General Pesticide Resources 229

G.2	Recognition and Management of Pesticide Poisoning, Including Materials for HCPs	230
	G.2.1 Pesticides and Human Health Concerns	231
	G 2.1.1 Telephone Hotlines	231
	G.2.2 Publications for HCPs	232
	G.2.3 Internet Resources	235
	G.2.4 Additional Resources Related to HCP Training on Recognition and Management of Pesticide Poisoning	236
	G.2.5 Pesticides and Animal Health Concerns	237
G.3	Pesticide Toxicology (also see Section G.4 Pesticide Products)	237
	G.3.1 Publications	238
	G.3.2 Internet Data Resources	238
G.4	Pesticide Products	240
	G.4.1 Databases on Pesticide Products	240
	G.4.2 Pesticide Product Labels	241
	G.4.3 Material Safety Data Sheet (MSDS) Directories	241
	G.4.4 Pesticide Manufacturers	242
G.5	Pesticide Usage	242
	G.5.1 Data Sources	243
	G.5.2 Reviews of Pesticide Use Data Sources	244
G.6	Pesticide Safety and Health Information to Assist Workers and Employers	245
G.7	Farmworker Employment, Demographics, Cultural Issues, and Service Organizations	248
	G.7.1 Employment Issues	248
	G.7.2 Child Labor in Agriculture	248
	G.7.3 Farmworker Demographics	249
	G.7.4 Farmworker Cultural Issues	250
	G.7.5. Farmworker Service Organizations	251
G.8	Nonoccupational Exposure Issues (Homeowner, Schools, Vector Control, etc.)	252
	G.8.1 Public Consumer Information	252
	G.8.2 Pesticides in Schools	252
	G.8.3 Vector Control and Pest Eradication Programs	254
G.9	State PPSP Contact Information	255
G.10	Federal Agency Contact Information	257
	G.10.1 U.S. Department of Agriculture (USDA)	257
	G.10.2 U.S. Department of Education-Office of Migrant Education	258
	G.10.3 U.S. Department of Labor (DOL)	258
	G.10.4 U.S. Environmental Protection Agency (EPA)	258
	G.10.5 U.S. Department of Health and Human Services–Centers for Disease Control and Prevention (CDC)	259
	G.10.6 U.S. Department of Health and Human Services–Health Resources and Services Administration (HRSA)	260

G.10.7 U.S. Food and Drug Administration (FDA) 260
G.11 Agricultural Safety and Health (Other Than Pesticides) 260
G.11.1 EPA 260
G.11.2 NIOSH 260
Index 265

Abbreviations

AAPCC	American Association of Poison Control Centers
AFOP	Association of Farmworker Opportunity Programs
ALT	alanine aminotransferase
AMA	American Medical Association
ANR	Agriculture and Natural Resource
ANSI/ASAE	American National Standards Institute/American Society of Agricultural Engineers
AOEC	Association of Occupational and Environmental Clinics
APHIS	Animal Plant Health Inspection Service
ASPCA	American Society for the Prevention of Cruelty to Animals
AST	aspartate transaminase
ATSDR	Agency for Toxic Substances and Disease Registry
BLS	Bureau of Labor Statistics
BOC	U.S. Bureau of the Census
BPHC	Bureau of Primary Health Care
CCOHS	Canadian Center for Occupational Health and Safety
CDC	Centers for Disease Control and Prevention
CDPR	California Department of Pesticide Regulation
CFR	Code of Federal Regulations
CIRS	California Institute for Rural Studies
CME	Continuing medical education
CPS	Current Population Survey
CPSC	Consumer Product Safety Commission
CSREES	Cooperative State Research Education and Extension System
CSTE	Council of State and Territorial Epidemiologists
DA	Department of Agriculture
DDT	dichloro-diphenyl-trichloroethane
DEET	N, N-diethyl-m-toluamide
DHHS	U.S. Department of Health and Human Services
DO	doctor of osteopathy
DOL	U.S. Department of Labor
DOT	U.S. Department of Transportation
EMT	emergency medical technician
EPA	U.S. Environmental Protection Agency
EU	European Union
EUP	experimental use permit
FAA	Federal Aviation Administration
FAQ	frequently asked questions
FBI	Federal Bureau of Investigation
FDA	U.S. Food and Drug Administration
FFDCA	Federal Food, Drug, and Cosmetics Act
FGIS	Federal Grain Inspection Service
FIFRA	Federal Insecticide, Fungicide, and Rodenticide Act
FJF	Farmworker Justice Fund
FQPA	Food Quality Protection Act
GAO	U.S. Government Accountability Office (*formerly U.S. General Accounting Office*)
GIS	geographic information system
HCP	health care professional
HDD	hospital discharge data
HIPAA	Health Insurance Portability and Accountability Act of 1996
HRSA	Health Resources and Services Administration

Abbreviations

HSDB	Hazardous Substances Data Bank	NCFH	National Center for Farmworker Health
HSEES	Hazardous Substances Emergency Events Surveillance System	NCHS	National Center for Health Statistics
IARC	International Agency for Research on Cancer	NEETF	National Environmental Education & Training Foundation
IATP	Institute for Agriculture and Trade Policy	NIH	National Institutes of Health
ICD	international classification of disease codes	NIOSH	National Institute for Occupational Safety and Health
ICU	intensive care unit	NPHSS	National Public Health Surveillance System
ILO	International Labor Organization	NPIC	National Pesticide Information Center
IPCS	International Programme on Chemical Safety	NPMMP	National Pesticide Medical Monitoring Program
IPM	integrated pest management	NSCEP	National Service Center for Environmental Publications
IRIS	Integrated Risk Information System	NTP	National Toxicology Program
LAN	local area network	NTSB	National Transportation Safety Board
LDH	lactate dehydrogenase	NUBC	National Uniform Billing Committee
LSC	Legal Services Corporation	NYDEC	New York Department of Environmental Conservation
MD	medical doctor	ODA	Oregon Department of Agriculture
MCN	Migrant Clinicians Network	OEHHA	Office of Environmental Health Hazard Assessment
MHP	Migrant Health Program, Bureau of Primary Health Care	OHSU	Oregon Health Sciences University
MMWR	Morbidity and Mortality Weekly Report	OPP	U.S. EPA Office of Pesticide Programs
MSDS	material safety data sheet	OSHA	Occupational Safety and Health Administration
MSPA	Migrant and Seasonal Agricultural Worker Protection Act	OSH Act	Occupational Safety and Health Act
NAICS	North American Industry Classification System	OSU	Oregon State University
NAIN	National Antimicrobial Information Network	PA	physician's assistant
NAS	National Academy of Sciences	PANNA	Pesticide Action Network North America
NASS	National Agricultural Statistics Service	PAPR	powered air purifying respirator
NAWS	National Agricultural Worker Survey		
NCEH	National Center for Environmental Health		

Abbreviations

PARC	Pesticide Analytical and Response Center	SENSOR	Sentinel Event Notification System for Occupational Risks
PCC	poison control center	SOC	standard occupational classification
PHI	protected health information	SPIDER	SENSOR Pesticide Incident Data Entry and Reporting
PIRT	Pesticide Incident Reporting and Tracking Review Panel	SPPC	SENSOR Pesticide Poisoning California
PISP	Pesticide Illness Surveillance Program	TBTO	Bis(tributyltin)oxide
PPE	personal protective equipment	TESS	Toxic Exposure Surveillance System
PPIS	Pesticide Product Information System	UB	uniform bill
PPSP	Pesticide Poisoning Surveillance Program	UNEP	United Nations Environment Programme
PVC	polyvinyl chloride	USDA	U.S. Department of Agriculture
RCW	Revised Code of Washington	WHO	World Health Organization
REDs	Reregistration Eligibility Decision Documents	WHS	Worker Safety & Health
RN	registered nurse	WSDA	Washington State Department of Agriculture
SAS	statistical analysis software	WSDOH	Washington State Department of Health
SDWA	Safe Drinking Water Act	WPS	Worker Protection Standard

Acknowledgments

The authors gratefully appreciate the listed contributors who supplied materials that were modified for this guide or provided significant input on the mechanisms of managing a pesticide-related illness and injury surveillance program, including conducting case investigations and coordinating with partner government and nongovernmental agencies. The authors appreciate the many other unnamed persons at State and Federal government agencies and nongovernmental organizations who contributed to this project by providing information. The authors also thank Diana L. Ordin, M.D., M.P.H., who originally conceived this project and Lorraine L. Cameron, M.P.H., Ph.D., who provided guidance in its early stages. The authors also gratefully acknowledge Jerry Blondell, Gene Harrington, Amy Liebman, Ray McAllister, Michael O'Malley, Carol Rubin, Patricia Schnitzer, and Michael Sprinker who reviewed an earlier version of this guide. Countless others have contributed by asking questions over the years, which we hope are answered by this document. Finally, creation of this document was made possible by contracts to Strategic Options Consulting, Inc from both the Oregon Health Division using funds obtained from the National Institute for Occupational Safety and Health (NIOSH) under Cooperative Agreement U60/CCU008161, and from NIOSH using funding support provided by the US Environmental Protection Agency.

Authors

Margot Barnett, M.S., Strategic Options Consulting, Inc.

Geoffrey M. Calvert, M.D., M.P.H., NIOSH, Centers for Disease Control and Prevention

Contributors

Lynden Baum, Washington State Department of Health

Michael Heumann, M.P.H., M.A., Oregon Department of Human Services - Health Services

Louise N. Mehler, M.D., California Environmental Protection Agency

Dorilee Peryea Male, New York State Department of Health

Rachel Rosales, M.S.H.P., formerly with the Texas Department of Health

Robert Stone, Ph.D., New York State Department of Health

Patrice Sutton, M.P.H., California Department of Health Services

Catherine Thomsen, M.P.H., Oregon Department of Human Services - Health Services

Editing

Jane Weber, M.Ed., NIOSH, Centers for Disease Control and Prevention

George Taylor, Cyrano

Design and Desktop Publishing

Donna Pfirman, AAS, NIOSH, Centers for Disease Control and Prevention

Vanessa Becks, NIOSH, Centers for Disease Control and Prevention

Printing

Patricia L. Ulakovic, NIOSH, Centers for Disease Control and Prevention

Web Production

Donna Pfirman, AAS, NIOSH, Centers for Disease Control and Prevention

David Wall, NIOSH, Centers for Disease Control and Prevention

About the cover

The cover drawing is titled "Path to Wellness" and was created by Blake Kidney. His was the winning entry in a Winter 2002 competition held at the School of Art, in the College of Design, Architecture, Art, and Planning, University of Cincinnati. The competition was open to undergraduate and graduate students at the school who were asked to submit works of art for the cover of this document.

1. Introduction

1. Introduction

1.1 A Guide to the Manual

1.1.1 Goal

This manual will help State health departments develop and maintain surveillance programs for acute and subacute health effects from pesticide exposure. It provides guidelines for program development, case investigation, data collection, outreach, and education. A list of resources for further information is also provided. The manual (which provides the case classification scheme, severity index, and sample data collection forms in the appendices), the standardized variables, and the SENSOR* pesticide incident data entry and reporting (SPIDER) computer program software described in the manual are intended to simplify and streamline the surveillance system development process. Adoption of these tools will allow a State to pool data with other State-based pesticide poisoning surveillance systems. Many tools and techniques covered in this manual may be generalized for surveillance of other occupational and environmental illnesses and injuries.

The manual is designed to address issues of capturing illnesses and injuries from pesticide exposures in workplace and nonworkplace settings. Pesticide poisoning is a complex condition for surveillance. It encompasses many illnesses and injuries created by single or mixed exposures to pesticide products. Pesticide products are often mixtures composed of pesticides and other ingredients that may have adverse human health impacts. The complex nature of pesticide poisoning and technical resources needed for case investigation warrants the development of surveillance programs based predominantly in State health departments or other State-level agencies.

1.1.2 How to Use the Manual

Surveillance of acute pesticide-related illness and injury requires a multidisciplinary approach that includes careful planning and implementation. This manual will be most useful when read in sequence—Chapter 1 through Chapter 9 and Appendix G—before implementing surveillance. Additional information that will be useful both in the initial phases of development and the ongoing implementation of the surveillance system is provided in the appendices. Readers working with established pesticide-related illness and injury surveillance programs can also use the manual to enhance their surveillance program and to find additional resources on pesticides.

1.1.3 Limits of the Manual

The surveillance system described here is not designed to address case and cluster reports of chronic health effects potentially associated with pesticide exposure (e.g., cancer, reproductive outcomes, or immunologic and neurologic effects of chronic exposure). While providing general guidance about parameters necessary for an effective pesticide poisoning surveillance system, it is not intended to cover every situation or to be a complete manual of standard operating procedures.

---

*SENSOR is the Sentineal Event Notification System for Occupational Risk.

1.2 Introduction to the Problem

1.2.1 Problem Overview

Over the past 20 years, concern about environmental health issues have increased, particularly in the area of pesticide exposure. These concerns have created a growing demand for health and environmental agencies to provide data on the impacts of pesticide exposure on human and environmental health.

Pesticides are toxic to certain life forms by design. In addition, they have the potential to cause adverse health impacts on humans and other nontarget species. The U.S. Environmental Protection Agency (EPA) has responsibility for ensuring that proper pesticide use does not pose unacceptable risks to humans and the environment. A variety of risk assessment tools are used to evaluate pesticide products, including both laboratory tests and field trials. The monitoring of acute illnesses associated with pesticide use is an important additional tool for identifying potential problems and populations at high risk, and to develop and evaluate risk reduction strategies.

1.2.1.1 The Changing Patterns of Pesticide Use

Pesticide use has expanded dramatically since the discovery of dichloro-diphenyl-trichloro-ethane (DDT) in 1939. Approximately 16,000 pesticide products are registered with the EPA. These products are based on approximately 600 active ingredients. Most pesticide products (approximately 80% by volume) are used by the agricultural industry [Donaldson et al. 2002]. In addition, a broad range of nonagricultural pesticide products are formulated for home, garden, structural, veterinary, antimicrobial, and insect repellent purposes.

Over time, the use of the most toxic pesticides has been decreasing as have the numbers of pesticide-related deaths and more severe poisonings. Advances in technology and the push for a more ecological approach to pest management will continue to shift the types of pesticides used over time. The current trend is toward greater use of biopesticides (microorganisms and pheromones) [NAS 2000]. Monitoring any adverse effects of these products on human and animal population health will be important. While these products currently represent only a small portion of the market, they are expected to play a larger role in the future. In addition, less toxic conventional pesticides will continue to be used.

1.2.1.2 Scope of the Problem

From 1993 through 1996, a total of 63,583 symptomatic poisonings from pesticides other than disinfectants were reported to the Toxic Exposure Surveillance System (TESS) [Calvert et al. 2001] and followed to determine a medical outcome. Of these, 16,258 (25%) were among children aged 6 and under. Workplace pesticide exposure accounted for 6,323 cases. According to TESS, an additional 22,889 poisonings were attributed to disinfectant exposures. These numbers exclude intentional poisonings such as malicious use or suicide. Suicides and misuse represent a relatively low proportion of reported TESS cases. The data from the TESS system are an indicator of the size of the problem of pesticide-related illness and injury in the United States. During the same time period, however, 3,143 occupationally related cases classified as definite probable or possible were reported to the California Department of Pesticide Regulation (CDPR) [CDPR 2002], suggesting that TESS may underestimate occupational illnesses by several fold.

1.2.1.3 SUSCEPTIBILITY TO PESTICIDE POISONING

The terms *pesticide poisoning* and *pesticide-related illness* are used interchangeably throughout this manual. These terms refer to acute and subacute illness or injury resulting from pesticide exposure. Whether pesticide exposure produces health effects in humans depends on the agent, the exposure scenario, and individual susceptibility. Agent-specific factors include the inherent toxicity of the pesticide, the physical characteristics of the formulation, and the presence of other compounds (e.g., adjuvants, carriers, emulsifying agents). Relevant exposure scenario factors include the dose (concentration and amount), route of exposure, duration and frequency of exposure, environment (heat, humidity, protective equipment), and any concurrent exposure to other substances. Individual susceptibility is influenced by many factors including age, sex, genetic composition, diet, and general health (e.g., presence of pre-existing illness). Health effects may result from acute or chronic exposure to high or low levels of pesticide products. Pesticide exposure may result in a wide range of symptoms dependent on the factors mentioned above. Acute illness may be mild (e.g., headache, rash, or flu-like symptoms) or more severe, including serious systemic illness, third degree burns, neurologic effects, and, rarely, death.

Some persons may have increased susceptibility to acute pesticide poisoning. Pesticide poisoning can affect both children and adults, although children may be more susceptible because of differences in organ system function and body composition. In addition, children have behavior patterns that might increase exposure. Finally, persons with asthma or other respiratory disease may also be susceptible to exposure effects despite proper pesticide application.

Mild illnesses from pesticide exposure are frequently characterized by nonspecific signs and symptoms, mimicking flu and other common illnesses. Responses to exposure may also be due to the odor or other irritant properties of the pesticide products as opposed to actual systemic intoxication. Health effects may result from intentional misuse, unintentional exposures, or use according to the product label.

1.2.1.4 A PROBLEM THAT AFFECTS US ALL

The widespread use of pesticides means that all sectors of the population are at risk of exposure. Occupationally exposed persons are at risk from exposure to more concentrated forms of pesticide if they are involved in manufacturing, reformulation, mixing, loading, or applying pesticide products. Workers who handle pesticides or pesticide-treated products risk illness arising from either chronic low-level pesticide exposure or a single acute pesticide exposure. Persons exposed to pesticides in the residential environment may have prolonged exposures if a pesticide product is misapplied to the residence or its surroundings. Children having repeated contact with pesticide-treated surfaces and those who spend large amounts of time in a treated home environment also may receive a substantial dose compared with others in the same residence. In addition, pets may serve as sentinels of exposure in these situations.

1.2.2 WHY INVESTIGATE REPORTS OF PESTICIDE POISONING?

There are several reasons for addressing reports of pesticide poisoning. Although pesticide products go through an extensive battery of testing before marketing, the testing protocol does not address all environmental conditions, mixtures of chemicals, chronic exposure patterns, and host parameters that can be encountered.

Surveillance serves as an early warning system of any effects not detected by manufacturer testing. It can also identify pesticide problems caused by noncompliance with pesticide regulations.

Investigation may reveal a pattern of problems associated with a particular pesticide active ingredient or a product formulation. An investigation can determine whether a pesticide illness event arose despite use according to the pesticide label, whether it was because of a violation of label instructions, or whether the label instructions were unclear, confusing, or inaccurate. This information can be used to determine if the product was used inappropriately, or whether changes are needed in label instructions, product design, or types of personal protective equipment (PPE) necessary to prevent additional illnesses from occurring. Information gathered through investigation can be used to detect whether particular populations are at greater risk, or whether activities are associated with exposure and illness that can be modified to prevent illness.

It would be ideal to have all States conducting surveillance of pesticide poisoning. In an era of limited public health resources, however, each State must determine whether this condition is a priority for the public it serves. The decision to implement surveillance may be based on the types and quantities of pesticides used in the State for agricultural, urban, or structural pest control, or in the absence of actual pesticide use data, the prevalence of crop/agricultural or other activities associated with high pesticide use. Other local issues may also drive the need to answer questions about the potential impacts of pesticides on public health. As another option, surveillance for pesticide poisoning may be integrated into a broader poisoning surveillance system.

1.2.3 Response to the Problem

The Council of State and Territorial Epidemiologist (CSTE) recommends that acute pesticide-related illness and injury be placed under surveillance in all States [CSTE 1996]. Additionally, U.S. Government Accountability Office (GAO) reports from the last 10 years have highlighted the need for standardized surveillance of human illness associated with pesticide exposure [GAO 1994, 1999, 2000]. Finally, the American Medical Association (AMA) supports the need for improved pesticide poisoning surveillance [AMA 1997]. Thirty States have rules requiring some form of physician reporting of pesticide exposure and illness, although most of these States do not have a surveillance program to act on these reports. Nine States (Arizona, California, Florida, Louisiana, Michigan, New York, Oregon, Texas, and Washington) conduct more comprehensive case investigation and surveillance activities. States with existing surveillance systems use a variety of systems for collecting and categorizing data. One objective of this manual is to raise awareness of the importance of adopting standardized coding and categorization systems. This manual will help States initiate a comprehensive Pesticide Poisoning Surveillance Program (PPSP). States considering developing a program may wish to follow the stepwise approach shown in Figure 1.1.

The EPA collects information about pesticide poisonings by a variety of mechanisms. It receives mandated reports of adverse effects from manufacturers, and periodically reviews both the TESS data maintained by the American Association of Poison Control Centers (AAPCC) and aggregated data from State-based surveillance programs. Additionally, the EPA receives more timely reports of significant illnesses and injuries from

State surveillance programs, State regulatory programs, and/or affected persons. State-based surveillance program reporting of significant pesticide-related illnesses and injuries to the EPA is voluntary but is viewed by the participating States as an important part of exchanging information to enhance the understanding and prevention of pesticide poisonings.

The National Institute for Occupational Safety and Health (NIOSH) has an interest in developing information about occupational pesticide-related illness and injury that will lead to prevention. It has provided funding to States for the development and enhancement of pesticide-related illness and injury surveillance programs. Most States with PPSPs report aggregated data to NIOSH that are shared with EPA and the National Center for Environmental Health (NCEH) at the Centers for Disease Control and Prevention (CDC). States may also request investigation assistance from NIOSH for particular types of cases (death, multiple affected persons, incidents involving new pesticide products, and incidents that occur despite use according to the product label). This cooperation helps provide a broader view of the problem of pesticide poisoning, and participating States benefit from the knowledge gained from pooled information.

Chapter 1 ■ Introduction

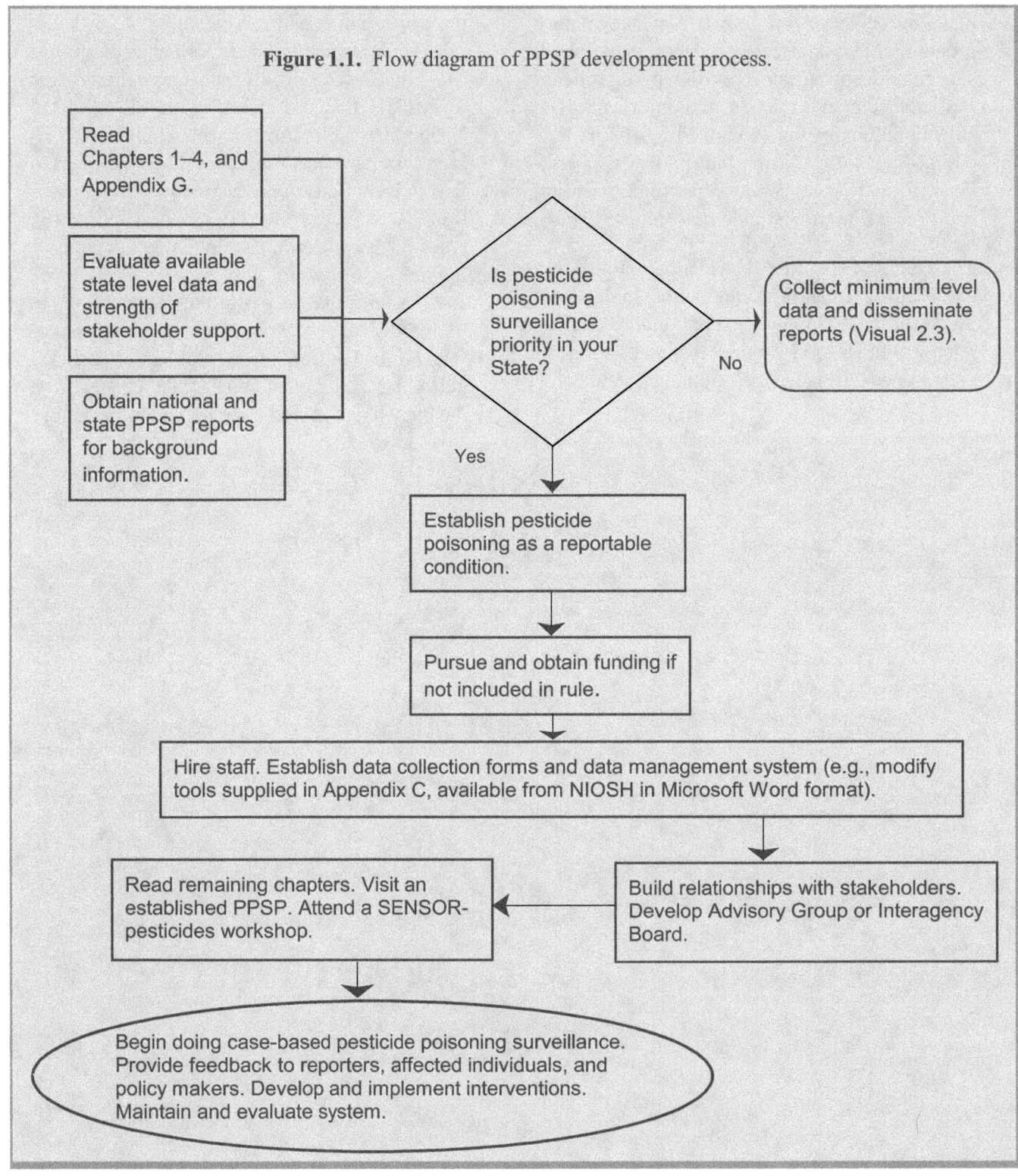

Figure 1.1. Flow diagram of PPSP development process.

2. First Steps in Surveillance System Design: Objectives, Resources, and the Reporting Rule

2. First Steps in Surveillance System Design: Objectives, Resources, and the Reporting Rule

2.1 Introduction

The design of the surveillance system should be based on the objectives of the surveillance program, the overall goals of the program's parent agency, and the level of resources available to conduct surveillance. Although this manual assumes that the surveillance program will be in the State agency having jurisdiction over health, we recognize that the surveillance program may be placed in a Department of Labor (DOL) or an equivalent agency. This discussion of surveillance system design is general and cannot address the various configurations that exist in different State governmental structures. It assumes that the reader is formally trained with a firm grounding in epidemiology and the general design of disease and injury surveillance systems. Many good resources are available to review the basic principles and practice of surveillance, and the information in those resources is not reproduced here [Teutsch and Churchill 2000; CDC 2001; Maizlish 2000]. The surveillance design described here includes passive case reporting mechanisms coupled with an active case investigation process. Many areas discussed are useful for a State that chooses to develop a surveillance system or to implement short-term surveillance projects. Some options are also provided for States without sufficient resources to conduct full-scale surveillance.

2.2 Objectives of Pesticide Poisoning Surveillance

The primary purposes of pesticide poisoning surveillance are as follows:

- Reduce the incidence of acute pesticide-related illness/injury.
- Identify clusters/outbreaks of pesticide-related illness/injury.
- Identify new pesticide problems and research needs.
- Identify high-risk pesticide active ingredients and products associated with pesticide-related illness.
- Identify groups at risk for pesticide-related illness.
- Document the distribution of acute pesticide-related illness.
- Target regulatory, enforcement, consultative, or educational interventions to prevent and control pesticide-related illness/injury.
- Evaluate the effectiveness of prevention efforts.
- Focus public attention on occupational/environmental health problems.
- Explore the feasibility of generating useful rate estimates and trend data.
- Generate research hypotheses.

At the individual case report level, the surveillance program may also assist health care professionals (HCPs) evaluate the patient's exposure situation and link the HCP with additional resources to help determine the patient's diagnosis.

2.3 Program Staffing and Structure

Running an effective PPSP requires a number of professional skills. The mixture of professionals who meet the needs of the program varies among the existing programs. Some programs have sufficient resources to maintain a full-time multi-member staff that includes program managers, data managers, case investigators, and field staff. Others are staffed more frugally with staff wearing multiple hats or split between various program activities, only one of which is pesticide poisoning surveillance.

2.3.1 Types of Expertise Needed for Surveillance

Surveillance for pesticide-related illness and injury requires program staff to have knowledge in a broad range of areas including the following:

- Toxicology
- Epidemiology
- Medicine
- Data management
- Occupational/environmental health
- Industrial hygiene

Other areas that are important but may be incorporated into the program by collaboration with other organizations include integrated pest management (IPM) and health education. The most successful PPSPs employ persons with training in epidemiology and environmental or occupational health. Employing or contracting with persons who are bilingual and bicultural to conduct interviews and participate in investigations involving non-English speakers is extremely important for program effectiveness. In most regions of the country, this usually means someone who can speak Spanish. In some areas, it may mean the program needs access to an interviewer who speaks Hmong, Mayan dialects, Russian, or other languages.

2.3.2 Ways to Meet Needed Expertise with Minimal Resources and Staffing Levels

It would be ideal for a program to have one person in each of the six main disciplines listed in Section 2.3.1, but most programs acquire this expertise by developing collaborative relationships with partners from other programs or agencies. Some level of medical expertise is certainly critical for effective surveillance. Because of the complexity of pesticide poisoning, a surveillance program should have access to a clinical toxicologist or a toxicologist and a physician familiar with the condition to assist with case or outbreak investigations and case classification. Credibility of the surveillance program is enhanced if a physician is either on staff or affiliated with the program. This may mean the State epidemiologist takes an active role in the program. If a physician is not available within the agency to provide assistance, a contractual arrangement with a clinical toxicologist or emergency physician consultant at a local university or hospital is an alternative solution. The National Pesticide Medical Monitoring Program (NPMMP) (see Appendix G) can also provide assistance to PPSPs and reporting physicians. The NPMMP can be contacted through the National Pesticide Information Center (NPIC). Over time, as surveillance program staff become fully trained and familiar with the toxicology of common pesticide classes, the day-to-day need for clinical expertise may decrease, and the consulting physician will be called on less frequently. The poison control center (PCC) may serve as a close partner to the surveillance program, depending on the relationships established by the health department. The Agricultural Extension Service may also have toxicologists based at the State land grant university who are familiar with the toxicology of pesticides as well as other agriculturally related toxins.

It is not absolutely necessary that staff have *a priori* knowledge about pesticides, although it is certainly helpful. If staff have no knowledge specific to pesticides when hired, they will need to become familiar with the subject quickly. New staff should be encouraged to attend education programs on recognition and management of pesticide poisoning conducted by the surveillance program or another source. Program staff will also need to develop sufficient understanding of pesticide toxicology to conduct case investigations and participate in classification of cases.

If program staff do not have a public health background, an introductory epidemiology course is useful. The CDC has a tutorial program entitled *Surveillance in a Suitcase* [CDC 2000a] that provides a solid grounding in surveillance. It is available on the Internet and complements the book *Principles and Practice of Public Health Surveillance* [Teutsch and Churchill 2000].

2.4 Program Funding Options

Several funding strategies are used by States with PPSPs. In California, the surveillance program managed by CDPR is funded by a tax on pesticide sales. Additional surveillance activities funded through a cooperative agreement from NIOSH are conducted by the Occupational Health Program at the California Department of Health Services. The surveillance program in Washington State is funded with State general funds supplemented by funding from a NIOSH cooperative agreement. Other States are reliant on a low level of general fund money combined with cooperative agreement funds from NIOSH. At times, States have also received funding from EPA and NCEH to support PPSP activities. Programs reliant on Federal funding have limited budgets and staffing compared with programs supported by general funds or sales fees.

2.5 Reporting Requirements and Rules

In the United States, State legislatures possess the authority for requiring disease reporting, which they exercise by enacting laws and statutes. In some States, pesticide poisoning and other conditions are specifically mentioned in a disease-reporting statute. In many States, State and local agencies are, by statute, delegated the authority to enumerate the reportable health conditions. In such cases, adding a reportable condition is most often a rule change rather than a statutory change. This section discusses elements found in statutes and rules that are useful for creating and maintaining a successful PPSP.

Both local and national information about pesticide use and poisonings have been used for justification when developing the reporting rule. At least three U.S. Government Accountability Office (GAO) reports discuss pesticide poisoning [GAO 1994, 1999, 2000]. Many states have found that GAO reports, annual reports from existing surveillance programs, published annual review articles from the AAPCC, and State workers' compensation data are useful resources to support a reporting rule and pesticide poisoning surveillance. Additional information can be obtained from the State agency responsible for enforcing pesticide regulations, which can provide material about complaint investigations that involve human health concerns. The series of articles from the *Morbidity and Mortality Weekly Reports* (MMWRs) listed in Appendix A are helpful examples of the way in which surveillance systems have helped identify particular problems associated with pesticide use. State-level information about calls received by the NPIC also provides some useful background information. The annual reports are available on the NPIC Web site: http://NPIC.orst.edu/reports.htm. (A link to an example of a pesticide-related illness reporting rule and justification appears in Appendix B.)

States have justified their reporting rule by citing the number of workers with potential pesticide exposure. Background information about migrant and seasonal farmworkers is located in data from the National Agricultural Workers Survey (NAWS) conducted by the U.S. DOL (data can be accessed at the following Web site: http://www.dol.gov/asp/programs/agworker/naws.htm). State- and county-level census data on the number of workers by occupation are available at http://www.census.gov/hhes/www/occupation.html. These census data can be useful for determining the number of workers in occupations having potential pesticide exposure (e.g., farm workers, pest control occupations).

The case definition for reporting purposes is generally broad and does not require a high degree of clinical diagnostic certainty. This approach will increase the sensitivity of the surveillance system for capturing cases of acute pesticide-related illness and injury. Unlike many other reportable diseases and conditions, pesticide poisoning encompasses a broad range of exposure agents and related symptomatology. For most health care providers, the evaluation of pesticide exposure and illness is a rare event. To ensure that the HCP or other source of case reports does not exclude potential cases, often the language in the reporting rule makes clear that cases need not be confirmed to be reported. Many States require that any *suspected* or confirmed case of pesticide poisoning be reported.

The reporting statutes and/or rules from several States, including California, Florida, Missouri, New Jersey, New York, Oregon, Texas, and Washington, are on the Internet (see Appendix B).

There are not many major differences across State rules/statutes. The examples listed in Appendix B represent those containing the broad language that is discussed in this chapter.

All States have significant problems with underreporting. State statutes/rules differ in exactly who is required to report. In some States, it is the licensed physician attending the affected patient; in other States, it is any health care provider aware of a case or suspected case. Considering the problems with underreporting, the broader wording is most effective for capturing the largest number of reports.

The PCC serving the State is a critical reporting entity to include in the surveillance program. PCCs often are specifically mentioned in the reporting rule, either by using generalized wording that they can interpret as including them or by developing a memorandum of understanding between the PCC and the PPSP. Similarly, workers' compensation data (both accepted and denied claims) are an important source of data on occupational pesticide poisoning, and kindred efforts should be considered for gaining access to them.

When developing pesticide poisoning reporting rules, consider the following important questions discussed in this chapter:

- Who is required to report, since the range of reporters will affect the completeness of reporting and the complexity of the surveillance system?

- Does the health department have authority to investigate and conduct site inspections of occupational exposure cases?

- Should the rule include a penalty for failure to report?

- Do the agency's existing confidentiality rules provide adequate protections to affected persons?

- Do enforcement agencies that receive referrals have the same confidentiality rules to protect medical information and/or the identity of the affected person? If not, is there an alternative referral approach that can be used?

- How will clinicians, the general public, workers, employers, and other stakeholders be informed of the rule change? How will they be given tools for recognizing, managing, reporting, and preventing pesticide poisoning?

2.5.1 Elements of the Reporting Rule

This section provides information about the elements contained in an effective reporting rule. Each State has different requirements for these rules and must make decisions and use wording based on their specific needs.

2.5.1.1 What Is Reportable?

Pesticide poisoning is a term easily recognized by clinicians, but it may cause them to limit their thinking to frank acute poisonings, no matter how it is defined in a rule. The term *acute pesticide-related illness and injury* is a more accurate description of what should be reported. In the rule, States specify whether the program is aimed at capturing only acute or both acute and chronic illness and injury. All of the information in this manual is limited to the surveillance of acute pesticide-related illness and injury, but some States may have reasons for wishing to capture both. Indicating that both clinically suspected or confirmed cases should be reported encourages health care providers to report even if they are not sure of the diagnosis.

Pesticide poisoning or *pesticide-related illness and injury*, whichever term is used, should be defined. The definition can make it clear that acute systemic, opthalmologic, or dermatologic illness or injury resulting from inhalation, ingestion, dermal exposure, or ocular contact with a pesticide is reportable. It is also helpful to use and define the terms *case, suspected case*, and *pesticide*. The definition for *pesticide* is generally the legal definition used by the State program taken from the State pesticide use laws. States may choose to make it clear that effects include those caused by both active and inert ingredients, and may choose to include *adjuvants* (see Section 2.5.3). (Adjuvants are materials that are added to a pesticide formulation to improve or change properties such as deposition, persistence, or mixing ability. These materials, which may be added by the pesticide applicator before a pesticide product is applied, include wetting agents, spreaders, emulsifiers, foam suppressants, and dispersing agents.) Since clinicians and the public often equate pesticides only with insecticides, confusion can be prevented by adding a statement such as: "Pesticides include but are not limited to herbicides, insecticides, rodenticides, repellents, fumigants, fungicides, and wood treatment products." It is important that educational materials for reporters and the public include information about classes of pesticides that may not be perceived as pesticides (e.g., herbicides, disinfectants, and wood preservatives). This definition is also where the surveillance program should indicate whether it is including or excluding illness and injury resulting from exposure to disinfectants.

In the spirit of having a reporting rule with broad wording, States consider whether to specifically include disinfectants, which are considered pesticides and produce a similar number of poisoning cases as are produced by conventional pesticides. Some programs, especially those with limited resources, may not be able to track disinfectant-related cases. However, including disinfectants in a reporting rule will facilitate their surveillance when additional resources are secured.

Ideally, making the full spectrum of pesticide-related illness and injury reportable is preferable to limiting reporting to occupational or nonoccupational cases. If jurisdictional or other limitations on resources exist, limiting reporting to occupationally related cases may be useful. Occupational exposures are more likely to be ongoing and have the potential to involve more toxic chemicals. However, having a broad reporting rule often makes it easier to build bridges with the agricultural community and to gain its support for the surveillance program. When surveillance is limited to occupational cases only, it must be made clear to the agricultural community that this surveillance also includes nonagricultural occupationally related cases.

An example of broad wording to define what is reportable is "Report cases or suspected cases of acute pesticide-related illness and injury when there is a history of exposure and a temporally-related illness or injury (laboratory confirmation is not required). For reporting purposes, *pesticide poisoning* includes acute poisoning as well as any subacute illness or condition (dermatologic, ophthalmologic, or systemic) caused by, or suspected of being caused by, pesticide exposure."

The statute/rule either specifies what must be reported in detail (e.g., a listing of name, address, phone number, social security number, sex, date of birth, diagnosis, etc.), or specifies that all information requested on an agency reporting form must be supplied to the health department. If the statute or rule does not clearly describe the agency's access to additional medical information or medical records, requests for medical information may be denied by the HCP or health institution where the affected person was seen. Likewise, the parent agency of the surveillance program should determine whether it has authority to gather information from third parties (e.g., employers and pesticide applicators) during an investigation. Some States have secured this authority through a change of the statute or rule for reporting of pesticide poisoning.

It may be useful to consider requirements for pesticide use reporting at the same time that the illness reporting rule is being developed and the PPSP is being designed. (see Appendix G for information about pesticide use reporting rules and data.) This is considered hazard surveillance, as opposed to disease surveillance. Pesticide use reporting can provide information about when and where hazardous pesticides are used, which can guide intervention efforts. In addition, pesticide use reporting can provide useful denominator data. For each pesticide or pesticide class, rates of pesticide poisoning cases per pound used of the pesticide can be calculated. These analyses would allow the identification of pesticides that poison the largest number of people per pound used. The disadvantages of pesticide use reporting are the time and financial burdens placed on pesticide users who must report this data, and on the State agency responsible for enforcing the rule and processing the data.

2.5.1.2 Who Must Report?

Reporting rules are typically aimed at licensed health care providers or physicians and, in some States, laboratories. A broad statement that is inclusive of a wide range of reporters is desirable, if no legal reasons for limiting the language exist. Some States require reporting by school nurses or school administrators for schools without a nurse. This may be a useful requirement if a State is including nonoccupational poisonings in the surveillance system. It would not be advisable if the system is interested in capturing occupational cases only. Surveillance systems that capture only occupational cases may confront difficulties when responding to school-based pesticide exposure events. Such surveillance programs can address the concerns of teachers and clerical and maintenance staff who may be ill from a school-based exposure event. However, the program's inability to address the public health concerns of students and their parents will create significant policy problems.

The PCC may be mentioned specifically as a reporting entity, or PCC staff may consider themselves to be health care providers under a broadly stated rule. This issue should be discussed directly with the PCC(s) in the State before developing language for a proposed rule. Similarly, workers' compensation data are an important source of cases, and kindred efforts should be considered for gaining access to it.

If reporting is mandatory, the State may choose to attach penalties for failure to report. This particular issue is often not directly addressed but should be considered. The disadvantage of penalties is that they may set a hostile tone. A clearly stated penalty may create a negative relationship with potential reporters when the State attempts to establish the reporting rule. The Washington statute includes a statement that no action shall result from the failure to report as required by the law, although it does allow the department of health to submit information about nonreporting primary care providers to the applicable disciplining authority [RCW 70.104.055(5)–(6)][†]. The California law contains a penalty clause that has been used very rarely to address a health care provider's failure to report. Washington originally proposed a similar clause in their law but changed it to the current wording after representatives of the State medical association made it clear they would not support penalties for failure to report [Baum 2001a].

2.5.1.3 Where to Report

The reporting process is usually standardized for all reportable conditions in a State with the report going to either the State or local health agency. It is easier and will prevent delays if reports go to the agency that will be conducting the investigation rather than to an agency that will only act as a filter or referral center. If reports go directly by the local health department, clear guidelines are needed to ensure reports are transferred to the State PPSP in a timely manner.

Some States stipulate that reporting may be to the Department of Agriculture (DA), the Department of Environment, or some other agency. For example, in Louisiana, reports go to the DA and the Department of Forestry. If reports go to an agency other than a local or State health department, it is critical that laws and rules ensure the appropriate level of medical confidentiality for reports and the portions of investigations that include medical information. (*Note:* Reporting rules requiring health care providers to report to a DA have not routinely resulted in health care provider reports. Most reports received by these systems come from affected persons complaining about pesticide applications made by another person.)

2.5.1.4 When to Report

Prompt reporting is critical if the surveillance program is designed to conduct timely investigations. A rapid reporting and response system permits information to be captured that might otherwise be lost, especially data available from environmental or biological specimens.

---

†Revised Code of Washington. See RCW in references.

By receiving reports promptly, the public health system can act to prevent additional exposures and illnesses. The range of reporting times in existing rules is from 24 hours to 30 days. Most States encourage telephone or faxed reports to ensure prompt reporting. Some States are moving toward electronic reporting: transmitting data in flat file ASCII or another standardized format has significant advantages in that it can be automated. Data are usually encrypted for security. This is particularly useful for reporters who have large numbers of reports or who provide batched periodic reports of data (e.g., laboratories, PCCs, or workers' compensation departments).

2.5.1.5 Health Insurance Portability and Accountability Rule and Public Health (HIPAA) Privacy Rule and Public Health Surveillance

The information in this section was adapted from the CDC publication entitled *HIPAA Privacy Rule and Public Health: Guidance from CDC and the U.S. Department of Health and Human Services* [CDC 2003]. This document is available on the Internet at http://www.cdc.gov/mmwr/preview/ mmwrhtml/su5201a1.htm.

New health information privacy standards have been issued by the U.S. Department of Health and Human Services (DHHS), pursuant to the HIPPA Act of 1996. The new regulations provide protection for the privacy of certain individually identifiable health data, referred to as protected health information (PHI). Balancing the protection of individual health information with the need to protect public health, the Privacy Rule expressly permits disclosures without individual authorization to public health authorities authorized by law to collect or receive the information for the purpose of preventing or controlling disease, injury, or disability, including but not limited to public health surveillance, investigation, and intervention [45 CFR 164.512(b)]. This includes the reporting of disease and injury for public health surveillance. A public health authority is broadly defined as including agencies or authorities of the United States, States (including public health departments and divisions), territories, American Indian tribes, or a person or entity acting under a grant of authority from such agencies and responsible for public health matters as part of an official mandate.

A public health authority at the Federal, tribal, State, or local level does not need disease or condition-specific laws before collection of PHI is authorized. On the contrary, public health authorities operate under broad mandates to protect the health of their constituent population, and they are authorized to receive PHI for the purpose of controlling disease, injury, or disability. A covered entity (that is, a health plan, health care clearinghouse, or health care provider who transmits any health information in electronic form in connection with a transaction [45 CFR 164.103]) may disclose the minimum necessary information to accomplish the intended public health purpose of the disclosure. The covered entity may rely on the public health authority's representation that the information is the minimum necessary to accomplish the intended public health purpose of the disclosure [45 CFR 164.512(b)].

To receive PHI for public health purposes, public health authorities should be prepared to verify their status and identity as public health authorities under the Privacy Rule. To verify its identity, an agency could provide any one of the following:

- If the request is made in person, the requestor presents an agency identification badge, other official credentials, or other proof of government status.

- If the request is in writing, the request is on the appropriate government letterhead.

- If the disclosure is to a person acting on behalf of a public health authority, a written statement that the person is acting under the government's authority is on appropriate government letterhead [45 CFR 164.514(h)(2)].

Public health authorities receiving information from covered entities as required or authorized by law [45 CFR 164.512(a) and 45 CFR 164.512(b)] are not business associates of the covered entities and therefore are not required to enter into business associate agreements. Public health authorities that are not covered entities are also not required to enter into business associate agreements with their public health partners and contractors. Also, after PHI is disclosed to a public health authority pursuant to the Privacy Rule, the public health authority (if it is not a covered entity) may maintain, use, and disclose the data consistent with the laws, regulations, and policies applicable to the public health authority.

Additional information about this topic appears in the CDC publication entitled *HIPAA Privacy Rule and Public Health: Guidance from CDC and the U.S. Department of Health and Human Services* [CDC 2003]. CDC recommends that public health authorities share the information in this document with health care providers and other covered entities and to work closely with those entities to ensure implementation of the rule consistent with its intent to protect privacy while permitting authorized public health activities to continue. Comprehensive DHHS guidance is located at the HIPAA Web site of the Office for Civil Rights http://www.hhs.gov/ocr/ hipaa/).

2.5.1.6 Confidentiality

It is assumed in this discussion that the State already has existing rules governing the confidentiality of personally identifiable medical information collected as part of disease reporting and special studies. This is an area that must be reviewed carefully if reporting is made to an agency other than the one that usually houses information about reportable conditions. For example, departments of labor, business services, or agriculture may not have adequate policies to protect confidential medical information. These issues may be addressed by carefully crafted regulatory language or a memorandum of understanding developed in consultation with the agency's legal counsel.

2.5.1.7 Interagency Cooperation or Sharing of Information

The mechanisms of interagency cooperation on investigations are discussed in Chapter 5. Some States have included statements about interagency cooperation in their laws or rules governing the reporting and investigation of pesticide poisoning (these statements may apply only to pesticide poisoning or apply to all reportable conditions or reportable occupational conditions). Several States have statutes and rules that specify the establishment of interagency boards related to the investigation of human illness associated with pesticide use. Oregon and Washington are two such States.

2.5.1.8 Authority to Investigate

In some States, the health department does not have clearly authorized access to workplaces unless they are establishments that are accessible to the broader public (e.g., retail establishments, schools, etc.). This is something that should at least be reviewed and considered when developing a statute and associated rules for surveillance of pesticide poisoning. To our knowledge, no pesticide poisoning rules exist that address the authority to conduct investigations. In contrast, some States have laws that address the authority to conduct investigations. Often, State health departments without a clear authority to investigate workplaces can gain access through voluntary cooperation. Employers are aware that failure to cooperate with an investigation will usually result in referral to an enforcement agency that has authority to investigate. (See Section 2.5.2.3 for further discussion of this issue.)

2.5.2 Examples of Reporting Laws and Rules

This section includes excerpts from laws and rules from the following States: New York, Texas, and Washington. These examples were selected for inclusion as they each contain elements that warrant some consideration for a State considering adding pesticide poisoning as a reportable condition.

2.5.2.1 New York Reporting Rule

This State's reporting rule (Visual 2.1) provides for reporting from health care providers and laboratories. It has clear statements about the reporting of cholinesterase analyses and other clinical laboratory testing for pesticides in human tissue. The wording is broad, requiring reports of confirmed and suspected cases. The requirement for laboratory reporting of cholinesterase results does contribute a significant number of reports that are unrelated to pesticide exposure. This is due to the routine evaluation of cholinesterase levels before administration of certain muscle relaxants used in surgery.

2.5.2.2 Washington Law

The Washington law [RCW 70.104 Pesticides—Health Hazards 2002] describing pesticide poisoning surveillance is more detailed than laws in most States. The definition of pesticide is very broad, specifically including spray adjuvants and agents intended to be used with pesticides. The statute includes language that empowers the Department of Health to "investigate all suspected human cases of pesticide poisoning and such cases of suspected pesticide poisoning of animals that may relate to human illness." The law also gives the Department of Health authority to take samples including human or animal tissue specimens for diagnostic purposes with the consent of the exposed person. This statutory provision permitting the department to obtain specimens appears in several other State laws. It is useful since it is very explicit and allows the specimens to be collected as part of the investigation to confirm the diagnosis. Without this explicit statement, States may find it more difficult to collect and analyze such specimens without a more research-oriented protocol; such a protocol may require institutional review board clearance and detailed informed consent. Note that in Texas, unlike Washington State, the statute empowers the health department to collect both biological and environmental specimens.

2.5.2.3 Texas Reporting Law

The Texas law contains a section (see Visual 2.2) on investigations that has a clearly stated right of entry authority for occupational cases, as well as the right to collect and analyze environmental and biological specimens. This wording provides access to the information needed to conduct complete investigations. Subsection (b) of the law might not permit inclusion of farm labor housing as part of an investigation. There may be interagency or constituency reasons why a State might choose not to include similar language in its law or statute. These issues should be explored before proposing language of this type.

2.6 Surveillance Strategy for States with Limited Resources

States with limited resources should consider adopting a completely passive system that uses existing PCC(s) data to report occupational pesticide-related injury and illness incidence as defined in Visual 2.3. This strategy does not require any active case follow-up or management of confidential information since data can be obtained without identifiers. Similarly, rates for nonoccupational pesticide-related illness and injury can be constructed by changing the demographic group and denominator. This surveillance approach does not require case follow-up, investigation, or a rule change. Other resource-sparing approaches discussed in this chapter include the following:

- Limiting the case definition to collect occupationally related cases

- Following up only on a subset of reports (e.g., severe illness, incidents involving multiple persons)

While these resource-sparing approaches provide an incomplete view of the problem of pesticide poisoning within a State, they do provide options for getting some sense of the scope of the problem, while using fewer resources than a more comprehensive surveillance program.

VISUAL 2.1. NYCRR TITLE 10, VOLUME A, PART 22 ENVIRONMENTAL DISEASES
(Statutory Authority: Public Health Law, §§ 225[5][t], 206[l][j])

22.11 REPORTING OF PESTICIDE POISONING. Every physician, health facility, and clinical laboratory in attendance on a person with confirmed or suspected pesticide poisoning or with and of the clinical laboratory results described in section 22.132 of this Part, shall report such occurrence to the State Commissioner of Health within 48 hours. This report shall be on such forms or in such manner as prescribed by the State Commissioner of Health.

Historical note
Sec. Filed August 14, 1990, effective August 29, 1990.

22.12 REPORTABLE LABORATORY TESTS FOR PESTICIDE POISONING. For the purposes of section 22.11, of this Part the following laboratory tests are reportable to the State Commissioner of Health:

(a) Blood cholinesterase levels that are below the normal range established by the clinical laboratory performing the test in accordance with quality assurance requirements established by the permit-issuing agency.

(b) Levels of pesticides in human tissue samples that exceed the normal range established in accordance with quality assurance requirements established by the permit-issuing agency.

Historical note
Sec. Filed August 14, 1990, effective August 29, 1990.

> ### Visual 2.2. Texas Reporting Law
> ### Communicable Disease Prevention and Control Act, Health and Safety Code, Chapter 84, The Occupational Condition Reporting Act
> (§ 84.007. Investigations)
>
> (a) The department shall investigate the causes of occupational conditions and methods of prevention.
>
> (b) In performing the commissioner's duty to prevent an occupational condition, the commissioner or the commissioner's designee may enter at reasonable times and inspect within reasonable limits all or any part of an area, structure, or conveyance, regardless of ownership, which is not used for private residential purposes.
>
> (c) Persons authorized to conduct investigations under this section may take samples of materials present on the premises, including samples of soil, water, air, unprocessed or processed foodstuffs, manufactured items of clothing, and household goods. If samples are taken, a corresponding sample shall be offered to the person in control of the premises for independent analysis.
>
> (d) Persons securing the required samples may reimburse or offer to reimburse the owner for the materials taken, but the reimbursement may not exceed the actual monetary loss sustained by the owner.
>
> _____
>
> Acts 1989, 71st Leg., ch. 678, § 1, eff. Sept. 1, 1989.
>
> Amended by Acts 1997, 75th Leg., ch. 245, § 6, eff. May 23, 1997.

Visual 2.3. Minimum Data Collection for Occupational Pesticide-Related Illness and Injury Surveillance

Below are guidelines for minimum data collection for occupational pesticide poisoning surveillance. Data should be obtained from poison control centers (PCCs) serving the State. Collecting these data will provide a State health agency with information about this condition that is comparable across States.

Data Resources	Poison Control Center data (numerator) BLS Current Population Survey Data (denominator) available at http://www.bls.gov/opub/gp/laugp.htm
Demographic Group	Employed persons aged 16 and older
Numerator	Reported cases of work-related pesticide poisoning defined as: 1. Exposure to an agent included in one of the pesticide generic categories (that is, fungicides, fumigants, herbicides, insecticides, repellents, or rodenticides), AND 2. Reason=occupational OR Exposure Site=workplace, AND 3. Medical Outcome is one of the following: minor effect; moderate effect; major effect; death; not followed, minimal clinical effects possible; or unable to follow, judged as a potentially toxic exposure.
Denominator	Employed persons aged 16 and older for the same calendar year
Measures of Frequency	Annual number of incident cases Annual incidence rate per 100,000 employed persons aged 16 or older
Time Period	Calendar year
Limitations of Indicator	Some States may not have a PCC. In addition, there may be rare circumstances in which a State health agency is unable to obtain data from their State-based PCC; however, under such circumstances it may be possible to obtain less timely PCC data from NIOSH at http://www.cdc.gov/niosh/pestsurv/.
Other Data to Collect from PCCs	Age, sex, pesticide active ingredient, signs/symptoms arising from the pesticide exposures, illness severity, and whether hospitalization/intensive care unit (ICU) treatment was provided.
Additional Guidance	Additional guidance on obtaining the numerator and denominator data are available from NIOSH (http://www.cdc.gov/niosh/topics/pesticides/) or from the Council of State and Territorial Epidemiologists (http://www.cste.org/pdffiles/Revised%20Indicators3.4.04.pdf).

3. Case Ascertainment

3. Case Ascertainment

3.1 Introduction

Several possible sources for pesticide poisoning case reports exist. Ideally, all of these sources should be used for timely identification of cases. However, if resources are limited, a single type of case ascertainment method may be chosen, supplemented by a periodic survey to review data from other sources.

3.2 Poison Control Centers (PCCs)

PCCs may function at a regional or statewide level. They receive calls from HCPs and the general public. The main function of PCCs is to provide toxicologic and case management information. Calls may be purely informational, but they commonly involve guidance on management of an acute ingestion or other acute exposure. PCCs often follow up cases until there is a final outcome, especially when there is a possibility that a person is at risk of more than minor adverse health effects. This follow-up information is used to determine the severity of the health effect. PCCs collect a variety of information including demographic data, the route of exposure, whether exposures were intentional, the site of exposure, case management, the therapy received, clinical effects by organ system, and medical outcome.

PCCs are an important source of case reports, especially for nonoccupational pesticide poisonings. As mentioned in Section 2.5.1.2, it may be helpful to list them specifically as reporters in the reporting statute or rules. The mechanisms and requirements for reporting should be discussed with the PCC prior to proposing language. Prompt reporting of cases by the PCC allows the surveillance program to act quickly to prevent additional exposures and illnesses from occurring.

The reporting guidelines shown in Visual 3.1 are useful according to State surveillance programs working with PCCs. Two data management software programs (Dotlab and TOXI-CALL®) commonly used by PCCs have developed customized reporting capabilities to facilitate reporting to PPSPs. These modifications include the capacity for real-time reporting to PPSPs. See Appendix C for instructions on obtaining case reports from PCCs, including a listing of the pesticide substance codes used by PCCs and information about search strategies for PCC data. PCCs can also assist reporting by physicians who call for advice on diagnosis and management of acute pesticide poisoning. The PCC can inform the HCP about the State reporting requirement and the PCC can offer to report the case. If the HCP agrees, the PCC may need to obtain additional patient information to satisfy the data reporting requirements (e.g., patient name and contact information).

Many PCCs have often struggled to maintain the funding required to remain open. In many States, PCCs receive financial support from the State department of health, which should facilitate the creation and maintenance of a reporting arrangement between the PPSP and the PCC. In States in which the department of health does not provide funding support to the PCC, the PPSP should consider making financial arrangements with the PCC. This will foster a

> **VISUAL 3.1. USEFUL REPORTING GUIDELINES FOR POISON CONTROL CENTERS (PCCs)**
> (Adapted from criteria used by Florida Department of Health)
>
> PCCs should report systemic pesticide poisonings (classic toxicosis) and those involving local responses (dermatitis, ocular effects, etc.) as well as reactions due to unpleasant pesticide formulation odors, pesticide product explosions, and allergic reactions. If an event consists of multiple cases, be sure to report information about each case. If Pesticide Poisoning Surveillance Program (PPSP) resources are limited, it may want to restrict PCC reports to the following cases involving pesticide exposures:
>
> 1. All occupational cases (that is, anyone with illness or injury associated with exposure to pesticides while he/she was at work):
>
>   - Including farmworkers, farmers, and pesticide handlers/applicators (pest control operators, golf course superintendents/technicians, pesticide manufacturing workers, etc. even when self-employed);
>
>   - Including office workers, teachers, construction workers, or persons employed in residential settings (home offices, residential service workers, etc.).
>
> 2. All serious cases, such as those resulting in death, hospitalization, or physician diagnosis of a poisoning (this includes attempted suicides).
>
> 3. All cases involving HCP-initiated calls in which the HCP describes clinical signs, or situations when callers are advised to seek medical attention. (Clinical signs can be systemic or local, including miosis, rash, conjunctivitis, dyspnea, etc.)
>
> 4. All cases, of any type, involving more than one person. This is intended to capture reports of mass poisonings in residential neighborhoods, schools, etc., where many people are reporting exposure-related illnesses due to a common source.
>
> 5. All cases involving exposure to public spraying of pesticides (e.g., medfly spraying, mosquito spraying, etc.), where the patient is symptomatic (even if there is only a suspicion that symptoms are related to the exposure).
>
> 6. Any other situation not covered here but considered eligible for a report by the PCC Director/Assistant Director.

stronger collaboration between the two agencies and will allow both agencies to better meet their obligations.

PCC data, stripped of individual identifying information, are aggregated into a national database by the AAPCC. This database (TESS) contains information about millions of poison exposures reported to certified PCCs in the United States. An annual report is published in the September issue of the *American Journal of the Emergency Medicine*. The annual report includes information about all toxic agents, not just pesticides. Additional AAPCC contact information appears in Appendix G. PCCs do not systematically collect detailed information about occupational cases (e.g., information is not collected on the worker's industry, occupation, or factors that led to the worker's exposure). Work-related information may be embedded in the narrative but is difficult to extract and is inconsistent when present.

3.3 Workers' Compensation Data

Workers' compensation claims can be a valuable source of information about occupational pesticide poisoning cases. States vary in coverage of agricultural workers by workers' compensation regulations (see Appendix G). In addition, thresholds for claim acceptance (that is, the level of documentation required, or type of illness/exposure) vary among the State workers' compensation systems.

The data collected by State workers' compensation programs vary widely, as does the accessibility of the data. States interpret the confidentiality of this information somewhat differently; therefore, access may be as simple as requesting a routine data transmission of the desired subset of variables, or may require development of a formal interagency agreement. If a surveillance program wishes to use workers' compensation data as a primary source of cases, this may require developing a formal agreement that allows the surveillance program early access to *submitted* claims data, including prompt access to hard copy or electronic physician reports. Including language in the reporting rule to permit access to the workers' compensation *submitted* claims data may be useful. Evaluation of these data on a monthly, quarterly, or annual basis will also permit a surveillance program to evaluate the completeness of reporting for occupational cases from other reporting sources. It may also provide information about a particular industry, demographic group, or type of exposure that is not reported through other sources.

It is preferable to obtain submitted claims data for both medical-only (these claims seek reimbursement of medical expenses only) and lost-time cases (claims that seek reimbursement for medical expenses and to recover lost wages). There are important reasons for gaining access to submitted claims versus *accepted* claims. The first reason involves timeliness. Workers' compensation claims are often submitted within hours or days of a pesticide exposure. However, it may be several weeks or months before the claim is accepted. Another issue is sensitivity. Although many submitted claims may be rejected, these rejected claims may meet the surveillance program's case definition for acute pesticide-related illness or injury. Access to submitted claims will allow the surveillance program to identify a larger proportion of the total universe of cases. One disadvantage is that some rejected claims are truly not cases of acute pesticide-related illness or injury. The surveillance program will expend some resources on following up on these claims that ultimately fail to meet the case definition. Visual 3.2 lists search strategies that some States have found useful when reviewing workers' compensation data. Additional approaches using nature of injury codes and international classification of disease codes (ICD) (e.g., ICD9 and ICD10 codes) may also be used, although this type of strategy is more useful when examining accepted claims data, due to the timing of when these codes are entered in the system. ICD9 and ICD10 codes that are useful for identifying pesticide poisoning cases are listed in Table 3.1.

3.4 Health Care Professionals (HCPs)

Physician (or, more broadly, HCP) reporting is the most common source of cases mentioned in reporting rules/statutes. While this method has been the mainstay of many communicable disease and notifiable condition reporting systems, it is not necessarily the most effective method for surveillance of pesticide poisoning. The nonspecific nature of symptoms arising from many pesticide exposures, difficulties of diagnosis, rare occurrence within an individual

Table 3.1. ICD–9 and ICD–10 codes to use when reviewing hospital discharge, emergency department, and workers' compensation data

ICD–9 Code	Condition*
989.0	Toxic effect of hydrocyanic acid and cyanides
989.1	Toxic effect of strychnine and salts
989.2	Toxic effect of chlorinated hydrocarbons
989.3	Toxic effect of organophosphate and carbamate
989.4	Toxic effect of other pesticides, not elsewhere classified
E861.4	Accidental poisoning by disinfectants
E863.0	Accidental poisoning by insecticides of organochlorine compounds
E863.1	Accidental poisoning by insecticides of organophosphorus compounds
E863.2	Accidental poisoning by carbamates
E863.3	Accidental poisoning by mixtures of insecticides
E863.4	Accidental poisoning by other and unspecified insecticides
E863.5	Accidental poisoning by herbicides
E863.6	Accidental poisoning by fungicides
E863.7	Accidental poisoning by rodenticides
E863.8	Accidental poisoning by fumigants
E863.9	Accidental poisoning by other and unspecified pesticides
E950.6	Suicide and self-inflicted poisoning by agricultural and horticultural chemical and pharmaceutical preparation other than plant foods and fertilizers
E980.7	Agricultural and horticultural chemical and pharmaceutical preparations other than plants, foods, and fertilizers
ICD–10 Code	
T60.0	Toxic effect of organophosphate and carbamate insecticides
T60.1	Toxic effect of halogenated insecticides
T60.2	Toxic effect of other insecticides
T60.3	Toxic effect of herbicides and fungicides
T60.4	Toxic effect of rodenticides
T60.8	Toxic effect of other pesticides
T60.9	Toxic effect of pesticide, unspecified
X48	Accidental poisoning by and exposure to pesticides
X68	Intentional self-poisoning by and exposure to pesticides
X–87	Assault by pesticides
Y–18	Poisoning by and exposure to pesticides

*Note: ICD–10 does not have specific codes for disinfectants. To find disinfectant poisonings, try T54, X49, X69, X86, and Y19, which are codes for corrosive and noxious substances. (Source: WHO [1977, 1992].)

> **VISUAL 3.2 USEFUL SEARCH STRATEGIES TO IDENTIFY PESTICIDE POISONING CASES FROM WORKERS' COMPENSATION DATA**
>
> In some States, the narrative portion (injured worker and/or physician statement[s]) of workers' compensation claim data may be searched using a computer; in others, the narrative is not entered into the data system.
>
> - For electronic searches, the following terms have been found useful: *cide, spray*, fumig*.
>
> - If physician narratives are screened, adding the terms organophosphate, *cholinesterase, 2-PAM, or atropine may yield additional cases.
>
> - If the data are being reviewed manually, additional search parameters include pesticide product names and all chemical exposures to agricultural workers, landscapers, maintenance workers, structural pest control operators, workers in pesticide and agricultural chemical manufacturing, and swimming pool service workers (this last occupation only if disinfectants are included in the surveillance system). Reports describing an agricultural worker with systemic or respiratory symptoms or a nonmechanically caused eye or skin injury should also be reviewed.

practice, lack of timely laboratory testing, selection of inappropriate tests, and reluctance to report cases make HCP reporting less reliable for this condition. Despite broadly worded reporting guidelines, HCPs are often reluctant to report cases for one or more reasons, including discomfort with reporting clinically unconfirmed cases, concern that an affected worker may experience job loss, perceptions that pesticide exposures are unlikely to cause illness, ignorance about the reporting requirement, and concern that reporting a case might disrupt any personal relationships with the employer.

All States with HCP-based reporting systems have conducted at least some level of HCP education to enhance reporting. Educational presentations on pesticide poisoning recognition and management provide HCPs with tools for recognizing the condition and understanding the reporting and case investigation process. Educational modalities include written case reporting guidelines, periodic case presentations in a health department or medical society publication, continuing medical education (CME) seminars (whole- or half-day), grand rounds presentations, tapes, videos, teleconferences, and Internet educational tools. As a mechanism for maintaining ongoing awareness that pesticide-related illness is a reportable condition, case vignettes and program updates can be included in a regular epidemiology publication sent to HCPs. Some combination of these different modalities can help maintain HCP awareness of the reporting requirements and astuteness in diagnosing potential cases. Any gains in HCP reporting associated with the implementation of these educational outreach efforts will be maintained only if the efforts are ongoing. Evaluation of educational programs can help a program fine tune their efforts. Evaluation tools include pre- and post-testing and examining whether attendees report cases within 1 year of training. Another approach is to compare the number of reports within a 3- or 6-month period after a large scale educational program, compared with the number of reports during the same time period in the previous year (comparing similar months will help account for seasonal variation in reporting).

Close linkages to a variety of expert resources are an additional enticement for HCP reporting. Providing contacts with clinical toxicologic expertise (e.g., through the local poison center, a university, the EPA, etc.), laboratory resources, or on-site sampling to help in the differential diagnosis can serve as an added incentive for reporting.

Selection of sentinel HCPs for more active reporting is a labor intensive process, yet may yield a number of cases that may not be identified through other reporting sources. The types of HCPs that are likely to yield the greatest number of cases include migrant health clinics, county health clinics, dermatologists, and emergency departments serving rural areas. Pesticide manufacturing or reformulation facilities may have contract medical staff who can also be contacted. Sentinel HCPs can be contacted to ascertain cases on a weekly or monthly basis, either in writing or via telephone.

3.5 Referral from Other Agencies

Other government agencies receive reports of pesticide-related illness and can be valuable sources for case finding. The obvious agencies include the following departments: agriculture, forestry, environmental quality, and the State Occupational Safety and Health Administration (OSHA) program. The number of case reports and validity of cases from these sources varies. Setting up good working relationships with the agencies and clearly defining the situations that warrant referral to the surveillance program are beneficial. A centralized emergency response program within the State, if there is one, can also be a source of case referrals. The regional EPA office sometimes receives complaints from the public, making it helpful to provide regional EPA staff with a description of the PPSP and guidelines for the types of reports that should be referred. Similar information can be provided to other Federal agencies with local jurisdiction that may be willing to refer cases, such as OSHA, the Department of Transportation, the Federal Railway Administration, and the Coast Guard.

Within the State health department, other programs with overlapping responsibilities for investigation may exist. Programs that are responsible for surveillance of hazardous substance spill or release events will usually also collect information about pesticide-related events. Drinking water and well testing programs, as well as indoor air quality programs, may receive complaints of human illness associated with pesticide exposure. It is important to develop mechanisms to coordinate with these programs.

3.6 Emergency Department Logs

Data are not recorded in any standardized fashion across hospital emergency departments and review of log information can be labor intensive. Despite their limitations, these data can be useful tools in developing or evaluating a pesticide poisoning surveillance system. Particular regional emergency departments may be useful as sentinel reporting sites. Periodic reviews of selected emergency departments' log data within a State, or smaller geographic area can be used to supplement surveillance data obtained from other case ascertainment methods. If emergency department records are available in electronic format, it may be useful to search these for the ICD9 and ICD10 codes provided in Table 3.1.

3.7 Affected Persons

More than half of the existing PPSPs accept initial reports from affected persons. The surveillance program often encourages these persons to seek medical attention. In some situations, the person may have already seen an HCP, but the HCP chose not to report. If this situation arises, the PPSP may choose to send a letter to the HCP with a reminder that pesticide-related illness and injury is a reportable condition. Appendix C contains a sample letter to address this situation.

3.8 Worker Representatives

Unions and legal services may function as referral organizations for persons, especially when the affected persons have concerns about confidentiality and potential retaliation from an employer or landlord. At times, contacts from these organizations may not provide sufficient identifying information for the health agency to conduct an investigation.

3.9 Hospital Discharge Data (HDD)

A set of extensive demographic, clinical, and financial information about every hospital inpatient is received by the hospital association, department of health, health care cost containment organization, insurance commission, or an equivalent organization in most States. This information is taken from the Uniform Bill 92 (UB–92), a document developed for use by third party payers and hospitals. The UB–92 Form (HCFA 1450) can be obtained from the Centers for Medicare and Medicaid Services' Web site http://cms.hhs.gov/forms/. Data elements are determined by the National Uniform Billing Committee (NUBC) convened by the American Hospital Association. This committee maintains the UB–92 data specification manual that provides detailed information about coding for the form. More information can be obtained directly from NUBC at http://www.nubc.org. The number of UB–92 data elements collected and used to create the HDD varies from State to State. Access to HDD is usually restricted by legislation. Agreements exist within each State about what elements of HDD are passed to State agencies involved in health policy and public health. The UB–92 includes a unique patient identifier for a person that can be used to track re-admission to the same or different hospitals over time to determine the course and outcome of injury. Unfortunately, there is frequently strict language in a statute or a memorandum of understanding that prohibits release of patient identifiers in the HDD abstract prepared for agencies.

The HDD abstract is usually made available on a quarterly or annual basis, which limits its use for timely case investigation. Health departments may have to pay for access to this data set. However, the HDD can be useful for determining whether the surveillance system is capturing the most severe cases of pesticide poisoning (that is, those requiring inpatient hospital care). Some States receive more timely HDD reports. For example, a revision to the New Jersey code for surveillance of hospitalized occupational and environmental conditions specifically requires reporting of notifiable occupational and environmental diseases and poisonings by hospitals using electronic HDD within 30 days of discharge. The rule also allows the program to request additional information in writing [New Jersey Department of Health 2000]. The same search strategy that is used for emergency logs or workers' compensation data, using ICD9 and ICD10 codes, can be used for HDD.

3.10 Laboratories

Clinical laboratories may collect specimens and conduct analyses for pesticides and metabolites in a variety of human or animal biologic media. The most common laboratory tests related to pesticide exposure are measurement of plasma pseudocholinesterase or red blood cell acetylcholinesterase levels, which are tests of cholinesterase inhibition. These tests may be conducted by hospital laboratories, local clinical laboratories, or referral laboratories. Other less frequently conducted tests include detection of pesticides (e.g., organophosphates) or their metabolites in blood or urine. In most cases, these other tests are conducted only by referral laboratories. Reporting rules vary by State about whether reporting is required from the physician ordering the test, the laboratory responsible for sample collection, or the laboratory conducting the test.

There are many complexities to interpretation of cholinesterase inhibition. A review of this topic appears in California's guidelines for monitoring workers exposed to cholinesterase-inhibiting pesticides [California EPA 2002], which are available at the following URL: http://www.oehha.org/pesticides/programs/Helpdocs1.html. Among the complexities is the wide normal range. Therefore, someone with a high normal baseline can have substantial cholinesterase inhibition and still have a level within the normal range. In addition, there are several different methods for conducting the tests, and all are subject to variability between and within laboratories. Cholinesterase tests may also be ordered to determine how a patient will respond to certain muscle relaxants used in surgery. This means that a depressed cholinesterase may be totally unrelated to pesticide exposure. One option, discussed by several States but not yet implemented, is requesting laboratories to indicate on the laboratory request form whether pesticide exposure is the reason for the test. This information would help surveillance programs and laboratories target resources toward pesticide-related laboratory test results.

The establishment of mandatory medical monitoring for workers exposed to cholinesterase-inhibiting pesticides coupled with a requirement for laboratory reporting is another approach that can be used. California and Washington are the only States that have mandatory requirements for such medical monitoring. The California Administrative Code, Title 3, Section 6728, requires medical supervision by a licensed physician for agricultural workers exposed to acute toxicity category 1 or 2 cholinesterase-inhibiting pesticides for 7 or more days in any 30-day period. Included with the code requirements is an extensive set of guidelines for physicians conducting medical supervision of these workers [California EPA 2002]. Washington State adopted a regulation effective in February 2004 that requires cholinesterase testing for some workers [Washington State Department of Labor and Industries 2003]. States considering laboratory reporting and/or requirements for medical monitoring of workers exposed to cholinesterase-inhibiting pesticides should review the findings of the California program [Ames et al. 1989]. An examination of this issue was conducted by an advisory committee in Washington [Washington State Department of Labor and Industries 1995].

4. Data Collection and Management

4. Data Collection and Management

4.1 Introduction

This chapter addresses some specifics of documenting PPSP procedures, including data collection and data management. Developing case investigation procedures, forms, and a data management system are important aspects of surveillance system design. (Case investigation procedures are covered in Chapters 5, 6, and 7.) If the appropriate information is not routinely collected, keyed, analyzed, interpreted, and disseminated, the goals of surveillance will not be realized. To some extent, the level of resources available to the program will dictate the amount of information that is routinely collected.

4.2 Data Standardization

At the outset, we cannot overemphasize the importance of using the standardized variables and the standardized case definition. Applying a standardized format for data collection makes aggregation of data across States feasible. The ability to aggregate data is valuable both at the State and Federal level. Potential users of the aggregated data include regulatory agencies, public health policy makers, researchers, programs conducting worker education, the public, and the medical community.

The large number of pesticide products on the market and difficulties in obtaining case reports makes the pooling of all available data particularly desirable. The ability to evaluate pesticide poisoning by product, crops, and geographic location greatly enhances the ability of States to evaluate whether limited case reports in their jurisdictions are reflecting larger problems linked to the specific uses of a pesticide product. Applications for this type of surveillance data at the national level cover a broad range of functions. Regulatory agencies can use the aggregated pesticide poisoning data to guide the development and amendment of regulations, target enforcement efforts, and evaluate the effectiveness of current control mechanisms. Pesticide users, commercial and agricultural pesticide applicators, and users of consumer pesticide products would all benefit from additional information that increases the understanding of risks associated with pesticide use.

4.3 Documentation of Procedures

The processes used to investigate cases, classify cases, enter and analyze data, provide feedback to reporters, and disseminate information are integral to a successful surveillance program. Documenting these procedures is often relegated to the bottom of the list of program management tasks. While programs can function without written documentation, it is certainly not advisable. Documentation provides guidance to all staff for a consistent approach to program objectives. Written policies and procedures make it easier to justify various decisions, including whether a particular pesticide exposure event will be investigated. As program procedures and policies change over time, written documentation facilitates the identification of those changes that might influence data analysis findings. Newly developed surveillance programs are often dependent on only one or two staff people who are working on many fronts. It is common for procedures to be developed and passed along verbally. Regrettably, if these staff leave the program, the procedures are often lost and must be recreated by new staff. For some procedures, a simple, bulleted list will suffice for documentation; for others, more complete instructions are desirable. The program should also maintain an orientation checklist for new staff to ensure that critical issues are covered in training. Topics of particular concern that should be documented include confidentiality, and safeguards to ensure employee safety and health when performing investigations.

Procedures should be developed for the following activities:

1. Case report management (intake, investigation, closure, classification, and feedback to reporters)

2. Data entry, quality assurance, and control

3. Data analysis

4. Data dissemination

Written policies and procedures should be developed for protecting the confidentiality of case report information at all stages of intake, investigation, and analysis. Examples of PPSP procedure documentation may be available through requests to established PPSPs. (Links to State program offices are in Appendix G under State PPSP Contact Information.) At later stages, policies and procedures for archiving data should also be implemented. These issues and the use of data to target activities and develop intervention strategies will be discussed in Chapters 5 and 6.

4.4 Data Collection

The PPSP collects data on each poisoning case, and these data are organized using variables. For all variables that are collected, States are encouraged to use standardized formats. Recommended formats are listed in *Standardized Variables for State Surveillance of Pesticide-related Illness and Injury*. (Copies of this document can be obtained from the NIOSH Web site, http://www.cdc.gov/niosh/topics/pesticides/, or by calling 1–800–365–4674.) The standardized variable format includes variable names, definitions, variable types, widths, and clarifying comments for variables that are considered desirable for all States to collect. Core variables that are critical for States to collect and transmit to CDC are indicated by asterisks. The variables are divided into general subject areas. Within any given subject area, variables are available that allow States to provide a brief narrative description about the data. Some additional discussion and clarification of the variables are provided below. The nature of the data collected for this condition usually dictates that States use a relational file structure and not a flat file structure.

4.4.1 Standard Variables to Be Collected by PPSPs

4.4.1.1 Administrative and Demographic Variables

The variables in this category include information about the source(s) of the report, relevant dates, event identifiers, county and State of exposure and residence, sex, age, Hispanic ethnicity, and race. These variables are used to describe the demographic characteristics of cases, track the geographic distribution of cases, and ensure that cases and events are linked without duplications. Not all of the variables needed at the State level are included in the standardized variable document (e.g., personal identifiers and addresses of the cases). However, these and other identifying and tracking variables are captured in the SPIDER database program. Note that race is not captured in the standard format currently recommended by CDC, since the CDC recommended format makes collection and analysis of race information more complex for persons who are multiracial. The race variable found in the standardized variable document is structured according to the CDC standard in effect at the time the document was initially developed and is considered easier to use by participating States.

4.4.1.2 Occupation and Industry Data

Coding of the occupation and industry of an affected individual can be accomplished by using one of several different standard coding systems. Codes for occupation can be based either on U.S. Bureau of Census (BOC) codes [NCHS 2003] or the 2000 Standard Occupational Classification (SOC) codes [OMB 2000]. Codes for industry can be based upon BOC codes [NCHS 2003] or North American Industry Classification System (NAICS) codes [OMB 1997]. Note that industry and occupation codes are periodically updated, with the BOC codes being revised every 10 years for use with the decennial census. The current BOC codes were used on the 2000 U.S. Census occupation and industry data, and these codes are referred to as the 2000 BOC codes. The 2000 BOC occupation and industry codes are based on, but are not identical to, the 2000 SOC and NAICS codes [OMB 2000], respectively. BOC codes are always 3-digit codes and therefore cannot provide the detailed industry coding provided by NAICS (which can code to 6 digits) nor the detailed occupation coding provided by SOC (which can also code to 6 digits). The NIOSH SENSOR-Pesticides Program recommends using the BOC codes for occupation and industry. This is because the number of workers in each of the BOC industry and occupation codes is available from Current Population Survey (CPS) data. CPS data can be used as the denominator to calculate rates of illness by industry and occupation. Although having all States use the same industry and occupation codes will facilitate the aggregation of data across States, States should choose the coding system that best suits their needs. Whichever coding system is chosen, States should use that system's most up-to-date codes.

Crosswalks are available to convert the NAICS codes into 2000 BOC industry codes, and to convert 2000 SOC codes into the 2000 BOC occupation codes. In order to convert industry and occupation data that may have been coded using older coding schemes, crosswalks are available to convert the 1990 BOC industry codes into the 2000 BOC codes and NAICS codes, and to convert the 1990 BOC occupation codes into the 2000 SOC codes and 2000 BOC occupation codes. All can be accessed at http://www.census.gov/hhes/www/ioindex.html.

Training on how to code occupation and industry is available periodically through the National Center for Health Statistics. This training usually covers all of the major occupation and industry coding systems.

4.4.1.3 Exposure Descriptions

The variables in this subject area help characterize the exposure. They describe the type of exposure (drift, direct spray, indoor air, contact, etc.), route(s) of exposure, whether the exposure was intentional, the person's activity at the time of exposure, and PPE worn by the exposed person. They also capture information about the equipment used to apply the pesticide, what the intended target of the application was, where the pesticide was being applied, and where the person was located when exposed (e.g., farm, nursery, home, school, manufacturing facility, etc.).

4.4.1.4 CHEMICAL INFORMATION

This section records information about the pesticide products associated with the exposed person's illness or injury. The system is not designed to capture information about non-pesticidal products such as fertilizers and *adjuvants*. Pesticide product information provided in SPIDER is adapted from the EPA Pesticide Product Information System (PPIS). This system can be accessed from the web at http://www.epa.gov/opppmsd1/PPISdata/index.htm. It is available in a searchable format on the Web site maintained by CDPR at http://www.cdpr.ca.gov.

States are strongly urged to collect sufficient data to permit full identification of the pesticide product whenever possible. However, at a minimum, pesticide functional class and product chemical class must be collected. This is in recognition that sometimes only minimal exposure information is available.

4.4.1.5 HEALTH EFFECTS DESCRIPTORS

This set of variables captures information about biological monitoring, medical diagnosis, pre-existing conditions, whether the person died, signs and symptoms, type of care received, and whether the person lost time from work or regular activities.

4.4.1.6 INVESTIGATION FINDINGS

These variables include enforcement agency findings, plus case investigation findings from the agency managing the surveillance program. Some variables are also specifically related to the Worker Protection Standard (WPS). The WPS variables address whether

- the incident involved re-entry into an area, field, or greenhouse treated with pesticide
- the worker had been informed of the re-entry interval for the treated area

(See Section 5.8.1 for more information about WPS.) In addition, a variable captures information about whether the product label was followed.

4.4.1.7 CASE CLASSIFICATION

These variables collect information about the components of the final case classification using the *Case Definition for Acute Pesticide-Related Illness and Injury Cases Reportable to the National Public Health Surveillance System* (NPHSS) described in Chapter 7 and provided in full in Appendix D. There is also a variable to record a separate case classification using either a separate State classification matrix, or to override the NIOSH classification matrix. For cases meeting the definition for reporting to the NPHSS, an additional component of case classification is a severity score of the illness/injury.

4.4.2 OPTIONAL VARIABLES

The variables in the standardized variable list that are not marked as core variables are all considered important, but are ones that some States may choose not to collect because of resource limitations. States are urged to collect as many standardized variables as possible. As already mentioned, the standardized variables include only those variables needed for national aggregation of data. Additional variables are needed for States to track and manage cases. Examples of some of these variables include personal identifiers; address and telephone number of the exposed person; name, address, and telephone number of HCP(s); laboratory sample tracking and results information for environmental and biological specimens; and information about animals (pets, livestock, and wildlife) affected by pesticide exposure. PPSP may want additional flags for particular types of cases that are of interest or concern at the

State level. Some but not all of these variables are captured in the SPIDER database.

The SPIDER system does not provide a tracking system to determine what information has been sent to or received from providers, individuals, and partner agencies. Developing a generic tracking system that would meet all States' needs is not feasible since the investigation and regulatory process is so different within each State. It is important for each PPSP to develop its own system for tracking cases. A tracking system can help to ensure that investigations are timely, that all necessary case information and medical or confidential information releases are obtained, that regulatory agency referrals and reports are received, and that appropriate feedback is given to relevant individuals, HCPs, employers, contract pesticide applicators, and partner agencies.

4.4.3 Introduction to the SPIDER Program

The SPIDER program is a data manager for collecting, managing, and reporting pesticide illness and injury data. Designed for NIOSH by the New York State Department of Health, the Program prepares data in the proper format for transmittal to NIOSH, and provides some pre-programmed reports used by PPSPs and NIOSH. The software was created using Microsoft Visual FoxPro, Version 5.0c, and Visual ProMatrix 5.0c. You do not need to purchase these products to run SPIDER [New York State Department of Health 1997]. Although additional reports can be created within SPIDER, more complex data analysis will require statistical analysis software (SAS).

System requirements for running SPIDER:

- Any IBM®-compatible computer with an 80486DX processor or higher.
- 32 MB RAM; 64 MB RAM recommended.
- A hard disk with 150 megabytes of free space. This will grow as cases are added.
- A 3 ½" floppy drive and a CD-ROM drive.
- VGA or higher resolution monitor running at 256 colors or more. SVGA (600 x 800) or XGA 1024 x 768) recommended.
- Microsoft Windows 95. (This is a minimum requirement. SPIDER also runs on more current systems, e.g., Windows 2000, NT4, and XP.)
- Installed Windows fonts: Arial, Courier New, and Times New Roman.
- A mouse is very helpful but not required.

This system can be installed on a local area network (LAN) for multiuser access. SPIDER is not equipped to upload cases reported electronically.

Other Options for a Surveillance Database

Some States have chosen not to use SPIDER, and have developed their own data systems for collecting information about pesticide-related illness and injury. If a State decides to develop its own database system and wants the ability to easily compare data with other States, and to contribute to national data, it is important to follow the standardized variable formats. It is equally important to contact NIOSH when developing a surveillance database to ensure that your system will readily transmit the necessary data in the desired format, and that you have the current version of the standardized variables. Some States have experienced problems incorporating chemical information into their databases in a way that will permit aggregation with data from other States. NIOSH may be able to provide assistance to ensure that chemical product data are collected and transmitted in a standardized fashion.

If a system other than SPIDER is used, the system should include documentation that describes the database, including a data dictionary, file structure, and table relationships. There should be written procedures for installation, operation, and maintenance of the system including how to backup the system.

4.5 Data Management

This section provides a brief overview of the elements needed for data management. It is provided as a reminder to new PPSPs that these are issues and elements that *must* be included in any data gathering program. For in-depth information about data quality assurance, data quality control staff should refer to the broad range of published literature and training programs available on this topic.

4.5.1 General Guidelines for Data Management

The importance of documenting surveillance procedures has already been emphasized. Procedures for entering reports into the data system, mechanisms to prevent duplicate entries, and management of discrepancies in information when a report is received from multiple sources should all be documented.

Having protocols for case triage and management, along with routine daily or weekly review of open cases will help ensure that data collection is complete and timely. Staff must be trained to have a clear understanding of the procedures and to strive for complete and accurate data. A clear procedure (e.g., a written protocol or an assigned coding administrator) should be in place to ensure that narrative data coding, interpretation of medical information, and pesticide product identification are performed in a consistent manner. There should be a system to monitor the quality of data entry to ensure the results comply with acceptable error rates. Staff should receive feedback on their data entry performance.

The SPIDER system contains many automated edit checks, as well as an audit trail, error reports, and missing data reports. If an alternative system is used, it should contain edits for missing data and errors (e.g., the program should identify codes that are outside acceptable ranges and illogical date sequences). Checks for duplicate records, blank records, orphaned data, and other anomalies created by changes in relationally linked data should also be part of the system.

4.5.1.1 Confidentiality and Security

The PPSP must develop systems for maintaining the confidentiality of hard copy and electronic records. Confidentiality procedures should be in writing to ensure that staff are clear about these procedures. Staff must understand the procedures and follow them routinely. Staff must also know whom to contact with questions. Records containing confidential information should be kept in locked file cabinets. Electronic systems should have passwords, and access to the system should be controlled.

4.5.1.2 System Backups

The administrator of the data program should establish written protocols for data system backup. Typically, there is a daily backup of data entered or edited during that working day; there should be a routine weekly or monthly backup as well. Safeguards for virus protections should be in place and routinely updated.

4.5.1.3 Transmitting Data to NIOSH

Annually, NIOSH assembles an aggregated database using data provided by participating PPSPs in the United States. In the past, NIOSH has requested that these data be provided by May 1. This gives the States a 4-month lag period from the end of the calender year to close out cases reported during the previous year. Grants and cooperative agreements awarded by NIOSH to fund PPSPs usually require this data sharing. PPSPs that receive no funding from CDC share their data voluntarily.

For SPIDER-using States, transmittal of data is relatively simple. These States need to ensure that their data are complete and then use the *Export to NIOSH* file command to prepare a zipped data file that can be transmitted to NIOSH. States not using SPIDER should contact NIOSH during development of their database system to discuss data transmittal issues. (Call 1–800–356–4674 or see http://www.cdc.gov/niosh/pestsurv/default.html.) All personal identifiers are stripped from the data before transmission to NIOSH.

Pooling surveillance data to create an aggregated database permits the creation of knowledge to prevent and control acute pesticide-related illness and injury. The aggregated database is shared with contributing PPSPs, NCEH, and EPA. Once data are checked for quality, accuracy, and absence of personal identifiers, it is made available for public access.

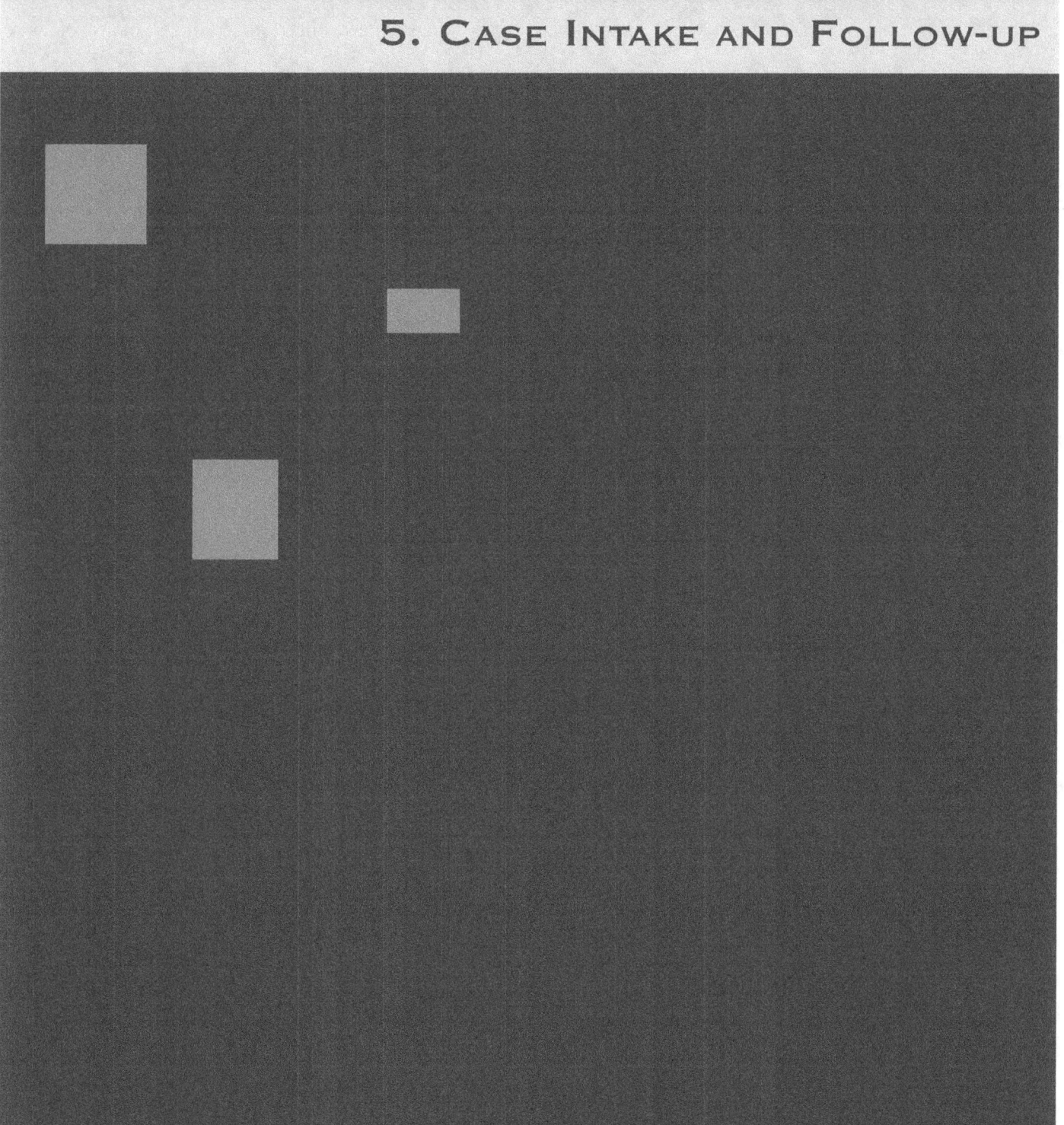

5. Case Intake and Follow-up

5. Case Intake and Follow-up

5.1 Introduction

This chapter pertains to surveillance systems whose case ascertainment relies on reports from HCPs, workers' compensation records, PCCs, referral agencies, affected persons, and laboratories. It describes some of the processes and issues associated with case report intake and follow-up. Investigative procedures for cases may differ slightly if long delays exist between time of exposure and receipt of the report, as is often the situation with cases identified via review of workers' compensation data. Timeliness of reports and the PPSP's response can impact the availability of persons for interviews, exposure site conditions, and the feasibility of sampling or collecting physical evidence. Extensive travel time for PPSP or enforcement agency staff to reach an exposure site can also have an impact on the amount of information that is available and therefore the outcome of a site investigation.

The guidelines provided in this chapter are designed for PPSPs without enforcement jurisdiction over pesticide manufacture, use, or disposal. In most States, another agency (e.g., agriculture) is charged with enforcement. These enforcement agencies have guidelines for identifying violations of pesticide statutes or rules. The inspection procedures and manuals used by these agencies are valuable references for nonenforcement investigators. Surveillance program staff are encouraged to use these manuals as reference guides. It is also helpful for surveillance program staff to accompany enforcement program staff on one or two inspections as an observer to gain a better understanding of the agency's inspection process. Any State initiating a PPSP is highly recommended to visit a State with an existing program, and to accompany the host State's staff on a site investigation, if possible.

5.2 Overview of the Case Investigation Process

The case investigation process includes all case-related activities beginning with case intake and ending with the case being prepared for case closure. The main goals of the case investigation process are as follows:

- Obtain sufficient follow-up information to determine whether the reported illness/injury meets the case definition of pesticide-related illness and injury.

- Provide information to the affected persons and/or their HCPs for case management and prevention.

- Provide prevention information and recommendations to the worksite (employer/workers) where the exposure event occurred.

- Determine if aspects of the exposure scenario require additional broader public health intervention.

- Disseminate information about the hazard and relevant prevention measures.

The level of action taken on each goal will depend on the chosen expertise and emphasis of the PPSP. Case follow-up includes the following

- Initial screening and triage of reports to determine whether they meet criteria for inclusion in the surveillance system

- Interviews with the affected person(s)

- Review of medical records, if available, and interview of HCPs, if needed

- Interviews with the applicator, employer, and/or owner

- Obtaining additional pesticide chemical information, as needed

- Identification of other exposed and affected persons

- Notification to the local health department, if necessary or required

- Notification to NIOSH and EPA, if necessary

- Referrals and interagency coordination for additional follow-up and investigation

- Final case review for completeness of data and collection of any additional missing data, if feasible

The initial case follow-up may be all that is needed to investigate a particular report. Other times, a report will meet the program's criteria for a site inspection by PPSP staff or a cooperating agency. Elements of the site inspection can include

- Environmental pesticide sampling

- Site evaluation

- Contacts with pesticide product manufacturers or equipment manufacturers

- Additional interviews

- Referrals and interagency coordination for additional follow-up and investigation

- Final case review for completeness of data and collection of any additional missing data, if feasible

- Regulatory action, if warranted, and/or recommendations for prevention

Case closure and classification involve:

- Evaluating whether information about the case is complete

- Assigning the case a classification category based on the standardized case definition

- Feedback to the reporter, HCP(s), affected person(s), and the worksite if appropriate

- Determining if the case warrants further efforts in terms of preventive intervention and dissemination of information

Note: Closure and classification may be provisional if a long time lag is expected for the final regulatory disposition of the investigation.

This chapter covers the case follow-up process. Site inspection, case closure, and classification are covered in subsequent chapters. The exact order of these steps may vary according to program protocols and the availability of information. A flow diagram of the case investigation process, similar to the example shown in Figure 5.1, can be helpful for program staff, and for explaining the process to partner agencies and the public.

5.3 Initial Report Intake (Complaint Evaluation)

This step in report management includes the collection of basic information about the affected person to determine whether the report meets criteria for additional investigation. This stage

Chapter 5 ■ Case Intake follow-up

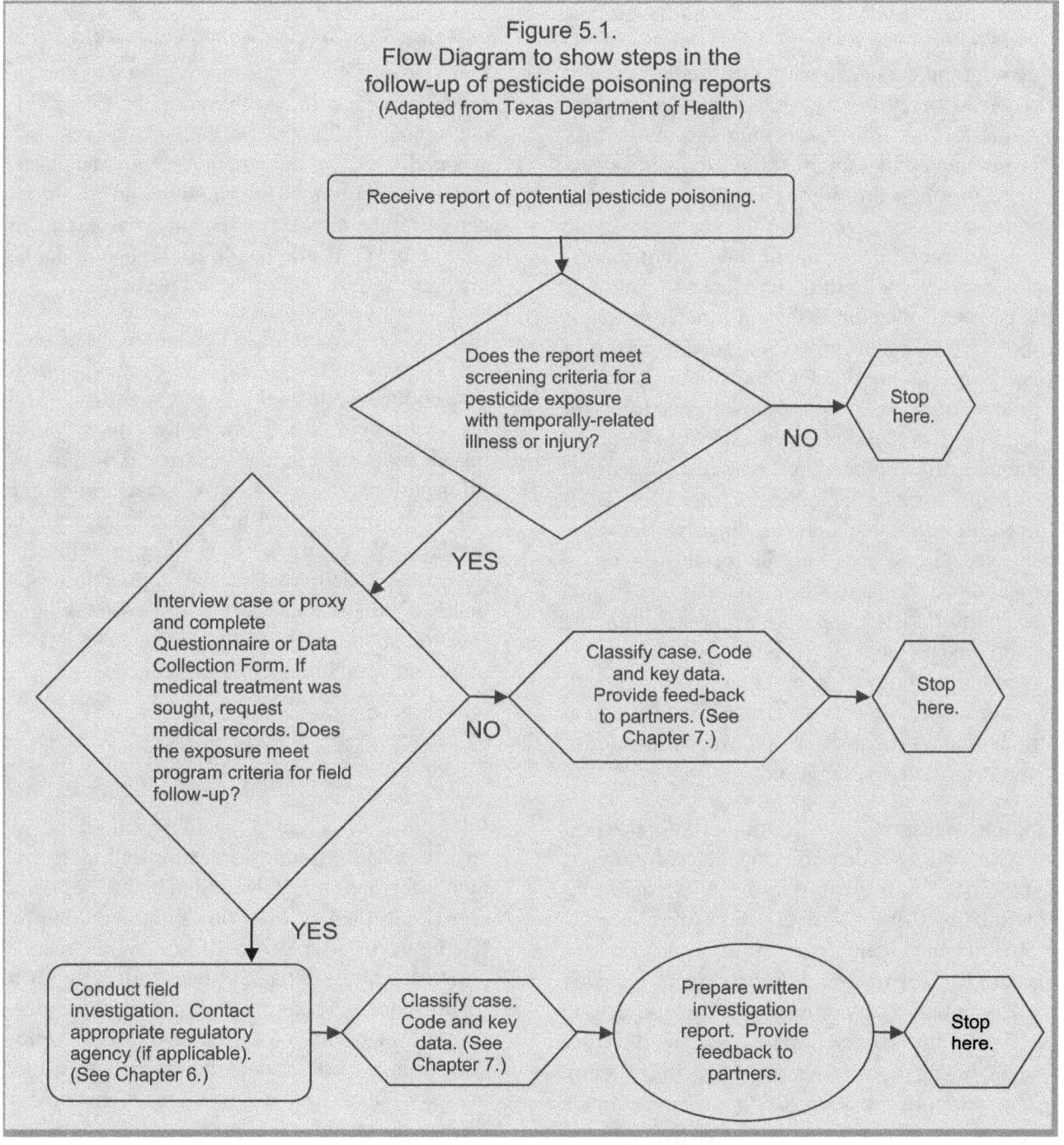

Figure 5.1.
Flow Diagram to show steps in the follow-up of pesticide poisoning reports
(Adapted from Texas Department of Health)

may include follow-up with the informant to determine if a pesticide exposure occurred and if temporally-related health effects developed. Depending on the source of the report, basic demographic and exposure information may be received in writing, by telephone, or in electronic format. Initial screening and intake may be conducted by support staff as long as a structured format is followed. The program administrator needs to determine if the program should log and track informational calls and/or reports that are screened out as unrelated to pesticide exposure. Collection of this information requires more work, but it provides a good measure of both service provided to the public and program workload. The step-wise procedures for logging in and assigning cases should be documented in a procedure manual. (A sample tracking form is included in Appendix C.) If multiple staff members are involved in case investigations, it is helpful for these staff to meet daily to ensure that individual reports that are part of a large exposure event are not being evaluated as separate events. Also, a weekly meeting to thoroughly review ongoing investigations will help provide structure to investigations, develop consistent program procedures, and prioritize investigations.

Simple questions asked during an initial report intake will help determine whether the person may have been exposed to a pesticide as defined by PPSP. These questions may seem so obvious that there is a risk they will not be asked for fear of insulting the person reporting the potential case. For example, early screening may exclude case reports by asking if symptoms began before the exposure of concern. The temporal relationship between pesticide exposure and symptom onset is critical.

Rapid identification of the chemical involved will allow staff to determine whether the chemicals fall within those covered by the PPSP. For example, a report may involve exposure to a disinfectant and these exposures might be excluded from the particular surveillance system. (Note: If a State does not collect these cases, the person can still be referred to the EPA product manager for the particular product. See Appendix G for information about databases that include product information and EPA contacts.) Other exposures that may be excluded from the PPSP are fertilizers, fire retardants, cleaning agents, and other nonpesticides.

It is always important to ask where the person was relative to the site of the suspect application about which they are concerned (e.g., did they actually see the plane or helicopter spraying versus just hear it in the area). This type of information may allow rapid screening of the call to determine what level of investigation is needed. For example, it is often possible to determine whether an aerial application was being conducted in the vicinity by making a few telephone calls, and if so, where it was done and what chemical was applied.

5.4 Case Follow-up Interviews

Interviews of affected persons should be as structured as possible to ensure that all pertinent information is collected in an efficient and consistent manner. Interviews may be done by telephone or in person. PPSPs should use a structured questionnaire or data collection form for all cases. Appendix C contains examples. Interviews of affected persons can be time-consuming. Staff should be trained to allow persons some time to voice concerns about their exposure but to control the interview and obtain the information needed to evaluate the exposure and illness. The interview should conclude with the interviewer summarizing the key data elements with the interviewee to be sure the interviewer has an accurate understanding of the events surrounding the exposure.

The person, HCP, or agency reporting illness should always be asked if there are additional affected persons. The PPSP may ask the index case or sentinel provider for contact information about other affected persons, or ask to have them contact the PPSP directly. This is important to ensure that all exposed ill persons are decontaminated and obtain needed medical care, as well as to ascertain the magnitude of the exposure incident. Cases may also be identified retrospectively and linked to a single exposure event by searching some of the data sources described earlier in Chapter 3. The procedures for following up on other affected persons may be governed by State rules about medical confidentiality. The level of effort expended to find additional cases must be weighed against the severity of illness, the likelihood of ongoing exposure, and the measures required to protect the confidentiality of the index case. Events that occur either in an unstable work environment (e.g., farmworker crew exposures) or involve a combination of public and worker exposures (e.g., retail establishments) require a prompt site evaluation to efficiently obtain information about additional cases beyond the index case. More stable workplaces can be followed up through telephone interviews.

5.4.1 Affected Persons

The affected person should be interviewed, whenever possible. The only exceptions are when the exposure is reported to be an intentional self-exposure (these interviews can be sensitive and are likely to produce little useful information) or the person is a minor. Interviews of minors should be conducted only with the permission of the parent or guardian.

The main purposes of these interviews are to

- Elicit information about the pesticide exposure and resulting illness or injury.

- Determine what factors caused the exposure.

- Provide information to help prevent ongoing or future exposures.

- Obtain HCP contact information if care was sought and this is a self-report.

- Ascertain if others are at risk.

PPSP staff should be aware that some nonoccupational case reports might involve pesticides exposures resulting from child neglect or abuse. The program should have guidelines for evaluating these situations and ensuring that appropriate agency referrals are made and properly documented.

It is important that PPSP staff be sensitive to concerns affected persons may have about possible repercussions from an investigation of their exposure. Renters may fear loss of housing if there is an investigation of a pesticide application made by a property owner. All occupational exposures may bring concerns about job loss. Immigrant workers may have additional concerns about immigration status and language difficulties understanding the investigation process.

Follow-up can be difficult if affected persons cannot be contacted by telephone. In-person interviews with such affected persons may be reasonable if the PPSP has regionally located field staff trained to investigate cases (affiliated with the State or local health department), or has a contractual relationship with locally based interviewers.

Note: If biological specimens are collected as part of the case investigation process, care must be taken to fully inform the affected persons what the specimens will be tested for and who will receive the results. This is particularly important since persons may be concerned that blood or urine specimens might be tested for drugs and alcohol, and that results could be given to employers or police.

When interviewing migrant workers, keep in mind that they might not remain in the area for an extended period. Those who are ill or injured may choose to leave the area and return to their stable home base for medical treatment; indeed, they are sometimes encouraged to leave by their employer or coworkers. It is helpful to get a permanent address for the migrant worker in order to inform them of the investigation results. Interviews involving occupational exposures should include evaluating the potential for take-home exposures (e.g., ask if the exposed person removes contaminated clothes and washes contaminated skin before returning home).

When interviewers are trained, it is important to emphasize issues of cultural sensitivity as well as proper techniques to avoid introducing bias into the interview process. These issues should also be considered when developing questionnaires. Many social science texts are available that contain guidance on interviewing techniques. One such text with an in-depth discussion of interviewing and the use of questionnaires from an anthropologic/ethnographic perspective is *Research Methods in Anthropology: Qualitative and Quantitative Approaches*, 3rd edition [Bernard 1995], particularly Chapter 9 "Unstructured and Semistructured Interviewing" and Chapter 10 "Structured Interviewing: Questionnaires." For a brief overview of epidemiologic issues associated with interviewing, see Hartge and Cahill [1998]. Appendix B of the Federal Insecticide Fungicide and Rodenticide Act (FIFRA) Inspection Manual also contains useful guidance on interviewing techniques [EPA 2002]. Since migrant farmworkers may feel more vulnerable because of concerns about job loss and deportation, resources on applicable cultural issues (including a link to a bibliography on farmworker living and working conditions) are provided in Appendix G.

When investigating a cluster of illnesses at a fixed worksite, school, residential institution, or a situation involving an exposed group of agricultural field workers, interview as many of the exposed persons as feasible. This will help to determine the range of symptoms, circumstances associated with symptoms, and circumstances that may have protected asymptomatic persons. If exposed persons received emergency decontamination or were transported to an emergency care facility, the investigator should also determine whether any of the emergency care providers were exposed and became symptomatic. If so, these persons should be included in case follow-up.

Be aware that some reports of pesticide-related illness may involve exposures and illnesses among large groups where the illnesses are ongoing and the source of exposure is unclear. (For example, a non-agricultrual workplace receives routine pesticide applications, is located near an agricultural operation, and a group of persons report ongoing or sporadic illnesses that are not associated with a specific pesticide application.) The investigative approach may be similar to that used in other noninfectious disease clusters. For instance, protocols for evaluation of indoor air complaints are particularly useful for these types of investigations [EPA/NIOSH 1991]. Staff should also be familiar with literature on epidemic psychogenic illness and ensure that the investigation process does not negatively influence the dynamic among the exposed persons [Alexander and Fedoruk 1986; Guidotti et al. 1987; Cole et al. 1990]. In some situations, it is useful to examine the incidence of symptoms in a control population of similar demographics. This approach has been used when investigating illness in school children and staff [Heumann 2000].

Particular care must be taken to accurately record identifying information when investigating clusters of illness among farmworkers or school children. Hyphenated last names and multiple names found in some cultures may result in duplicate cases, especially when the names are presented in differing orders. Also note that with hyphenated names, sometimes only one name will be recorded. Careful collection of name, age, date of birth, and addresses will help to avoid duplicate entry of cases reported from multiple sources.

5.4.1.1 Chemically Sensitive Persons

Chemically sensitive persons may ask the PPSP for recommendations to prevent pesticide exposure. In some States (e.g., Florida, Louisiana, and Washington), the DA maintains a registry of residents who require notification before pesticide applications are made near their homes. To be placed on the registry in Florida, a person must provide a note from his or her physician and must pay an initial registration fee and an annual renewal fee. Those on the registry are notified at least 24 hours before any relevant pesticide application in the vicinity of their property. Notification can be made by telephone, mail, hand delivery, or in person. The Florida statute appears at: http://www.flsenate.gov/welcome/index.cfm (see Title XXXII, Chapter 482, Section 2267). Given that these registries can help chemically sensitive persons avoid potential pesticide exposures, it is recommended that the PPSP determine if their State has such a registry. In States without a registry, some chemically sensitive persons have established pre-notification agreements. Pre-notification agreements are informal and involve a request to neighbors to provide sufficient notification before application of any relevant pesticides.

PPSPs will occasionally receive reports from chemically sensitive persons who claim they were poisoned by pesticides. If such a claim is for a substantial exposure, the person's complaint to the State DA may result in an investigation. These investigations often include sample collection to determine the presence of residual pesticide, which if detected may cause the DA to take action against the pesticide applicator (e.g., verbal or written warning for a first violation). Involvement of the DA can help to favorably change the behavior of the pesticide applicator, whether a neighbor or commercial operation.

5.4.2 HCPs

A telephone report received from an HCP's office can provide more information than a written or electronic report. However, a telephone report may be incomplete if the person is still symptomatic or undergoing testing and/or treatment. If the patient is still present at a clinician's office when the report is made, you have an excellent opportunity to provide assistance to the HCP, such as tracking down information about the pesticide product.

For reports received after the person's symptoms have resolved, follow-up interviews with HCPs should generally be made only when a review of medical records does not provide critical information needed to classify a case. During HCP follow-up interviews, provide opportunities for the clinician to ask questions or to provide insight into the patient's exposure. These interviews should be kept as brief as possible and be organized so repeated contacts will not be needed. follow-up calls can be irritating in a busy clinical setting, especially if they are caused by disorganized data collection. Before conducting a follow-up interview, be sure to highlight all questions with missing information on the data collection form (or questionnaire) to help ensure that you obtain all of the information needed. PPSP staff responsible for contacting HCPs should be trained in medical terminology and have access to reference materials on standard diagnostic tests.

Migrant and seasonal farmworkers are mobile and may not be easily accessible by telephone for follow-up. Clinics that serve this patient population should therefore be encouraged to report suspected cases while the patient is still in the office. It is equally important to have bilingual/bicultural staff or to contract with interviewers familiar with the worker populations in your area.

Most programs routinely obtain medical records of reported cases. Some States have indicated that having the medical record request signed by a health department physician is more likely to yield medical records than when letters are signed by staff who are not physicians. To protect patient confidentiality, information obtained from a medical record review must be carefully guarded to ensure it is not released to other agencies cooperating on an incident investigation.

5.4.3 Third Parties-Pesticide Applicators, Landlords, and Employers

Interviews with or written requests to third parties for information are commonly part of the case follow-up process. Third parties can include employers, pesticide applicators, and landlords, especially in the event of agricultural exposures involving drift or spray from aerial applications. An exposure scenario might involve several employers; for example, the employer of the pesticide applicator, another employer who contracted for a pesticide application, a third employer whose workers in a nearby field were subjected to drift, and possibly a labor contractor who is the actual employer of the exposed workers. Although third parties may not be legally required to cooperate with an investigation, they usually do.

Third-party interviews are often critical in determining exactly what pesticide the affected person was exposed to as well as additional factors that may have influenced the exposure. The employer or applicator can supply information about anything unusual about the application. The applicator should also be able to supply information about application equipment and methods, product dilution, mixtures of products, and any *adjuvants* added. When investigating illness in a crew of field workers or a stationary workplace, the employer should also be able to supply names and contact information of the exposed or potentially exposed workers.

If staff conducting interviews and site inspections have limited first-hand knowledge of particular types of pesticide applications, training them for interviewing pesticide applicators can be helpful. This could include simulated interviews of volunteer applicators using an unscripted case scenario to familiarize the interviewer with important information–namely, correct terminology, what the application equipment looks like, and what might go wrong with an application. It is helpful to ask these volunteer applicators about past problems they experienced or observed. Their responses and "war stories" can provide some clues for areas to home in on during an investigation. It is best if the interview is conducted at a site where the applicator has access to equipment and record-keeping forms to help familiarize the interviewer with these items. Members of the PPSP advisory committee (see Section 5.9.3) representing various sectors of the pesticide industry may be able to assist with identification of volunteer applicators for this interview process.

It is important for staff to be carefully trained to protect the confidentiality of affected persons to the extent possible. There are clearly times when complete anonymity of the affected person is not possible, especially when dealing with a small workplace or investigation of an application to a residence. If a person is suffering from mild illness and is likely to be easily identified if questions are asked (e.g., he is the only pesticide handler at the workplace), the PPSP must determine if obtaining the third-party information is critical. In these situations, it is important to gain the permission of the ill worker before contacting the employer. Staff should be familiar with protections against discharge and employment discrimination provided to workers by provisions of the Occupational Safety and Health Act [29 CFR‡ 1903.11c] (OSH Act) and other State regulations.

5.5 Notification of the Local Health Department

The relationship between the agency housing PPSP and the local health departments varies by State. In general, the minimum level of notification is that all cases received at the State level are reported within 24 hours to the local health department, and a brief final summary is provided when the investigation is complete. If reports involve multiple persons or a broader public exposure, the contact should usually be more in-depth.

In a number of States, local health department staff are trained to conduct pesticide illness investigations and may take the lead in some case investigations. Local health departments may become involved in investigations involving large numbers of exposed persons, and clusters where the cause of an illness cluster is unclear after a preliminary investigation. Even if local health department staff do not conduct full investigations (interviewing clinicians, exposed persons, etc.), they can be valuable members of the investigation team by collecting and transporting samples.

5.6 Obtaining Pesticide Product Information

Ideally, the exposed person, treating HCP, or another informant will be able to give the PPSP a product name and EPA registration number. More commonly, substantial sleuthing may be required to obtain this information. It is important to try to get both the product name and EPA registration number since some products with the same or very similar names have very different formulations. Each product has a unique EPA registration number, and this can be used to differentiate products with the same name. If the exposed person is not the person who applied the pesticide, the pesticide applicator will need to be contacted by the PPSP or a partner agency depending on the program protocols to obtain pesticide product information.

The SPIDER database, POISINDEX®, EPA PPIS database, and the PANNA Web sites are all good starting points to find information about a pesticide product's active ingredients. Links to some of these sites, and commercial sites for product labels and material safety data sheets (MSDSs) are provided in Appendix G. Note that these databases generally only contain information about active ingredients in the pesticide formulation (that is, those chemicals added for the purpose of their pesticidal activity) and not about inert ingredients.

---

‡*Code of Federal Regulations.* See CFR in references.

5.6.1 Inert Ingredients

Pesticide products may contain ingredients that are considered inert as defined by FIFRA. Inert ingredients are not included in the formulation for their pesticidal properties (although they may possess such activity). In 1987, the EPA developed a policy to "reduce the potential for adverse effects from the use of pesticide products containing toxic inert ingredients" [52 CFR 13305]. As part of this policy, inert ingredients were categorized into four lists based on hazard and priority for testing. All of the inerts on List 1 (categorized as chemicals of toxicological concern) must be listed on the product label (see Table 5.1). This categorization is based on carcinogenicity, neurologic effects, developmental and reproductive effects, or adverse ecological effects. Some pesticide products voluntarily indicate on the product label the identity of inerts or other ingredients not on List 1. The use of the term *inert* is accepted, although EPA now encourages registrants to use the term *other ingredient* rather than *inert* [EPA 1997]. This policy change is a result of EPA's efforts to make the language on pesticide product labels clearer for consumers. (See http://www.epa.gov/opptintr/labeling/ for information about the Consumer Labeling Initiative.)

List 2 provides the 95 chemicals currently used in pesticide products that are considered potentially toxic. These ingredients are undergoing review to determine whether they should be moved to List 1 or List 4, described below. This determination will be based on an assessment of carcinogenicity, neurologic effects, developmental and reproductive effects, or adverse ecological effects.

List 3 contains inert ingredient chemicals of unknown toxicity that are undergoing assessment. List 4 is composed of two lists: List 4A are minimal risk inert ingredients and List 4B are those inert ingredients that the EPA has determined to pose no adverse risks to the environment or public health when used in pesticide products. *Note*: Lists 2, 3, and 4 are not included here because of their length and more frequent updating compared with List 1. New inert ingredients are occasionally added to the lists, and the EPA issues periodic notices of reclassification for chemicals that have undergone review. Lists 2, 3, and 4 and additional information about inert ingredients appear at http://www.epa.gov/opprd001/inerts/. This Web site also has links to Federal Register notices that list inert ingredients removed from all lists but which may be found in older products associated with exposure incidents.

Regrettably, product-specific information about inert ingredients is not readily available to the public or public health professionals in any database. Several lawsuits have been filed in an attempt to require disclosure of inert ingredients on the product label, but the disclosure is currently voluntary except for inerts on EPA's List 1. The inert ingredients are considered *Confidential Business Information* protected under the trade secrecy provision of FIFRA. Product MSDSs may contain some information about inert ingredients. If a PPSP suspects that

Table 5.1. List 1: Inerts of toxicologic concern—currently used in pesticide products.

Chemical Name	CAS Number
1,4-Benzendiol	123–31–9
Diethylhexylphthalate	117–81–7
Dioctyl adipate	103–23–1
Ethylene glycol monoethyl ether	110–80–5
Isophorone	78–59–1
Nonylphenol	25154–52–3
Phenol	108–95–2

an inert ingredient may be implicated in an illness or injury, the product manufacturer or the local PCC should be contacted to obtain information about that ingredient. Contacting the manufacturer or the local PCC is usually faster than trying to obtain the information from the EPA. Always ask to speak with a manufacturer's toxicologist or physician when requesting this information. You may need to submit a request via fax to confirm the public health need for the information. Registrants are required to provide information about inert ingredients to HCPs involved in the evaluation of an exposed person. For links to manufacturer and MSDS Web sites, see Appendix G. Keep in mind that additional carriers or *adjuvants* that might have adverse toxicologic effects may be added by the pesticide applicator. These carriers and *adjuvants* are not registered as pesticides, although they are designed to be mixed with pesticide products. Information about carriers and *adjuvants* is available from some of the sources listed in Appendix G. The NPIC Web site (http://npic.orst.edu/manuf.htm) is an additional good source of contact information for manufacturers.

In the event that inert ingredient information is not available from either the manufacturer or the local PCC, this information can be provided by the EPA [40 CFR 2.307]. Currently, this requires the health department to place a request on its letterhead (or the letterhead of another government agency with responsibilities for protecting public health) and fax it to the Assistant Administrator of EPA's Office of Prevention, Pesticides, and Toxic Substances. The fax number is 202-564-0801.

Inert ingredients are more of an exposure concern for particular product formulation types. Use of aerosol products can result in exposure to active and inert ingredients due to the small droplets dispensed by the pressurized product.

The inert solvents and propellants in aerosol products can be hazardous. Liquid formulations (emulsifiable concentrates, soluble concentrates, liquids, ultralow-volume concentrates, and solutions) are of greatest concern. The inert ingredients can include oils, solvents, or alcohols, and concentrated (as opposed to ready-to-use) formulations may contain high levels of active ingredients. Some liquid formulations may contain antifreeze to prevent freezing in storage.

Pesticide dusts are composed of finely ground pesticide mixed with a dry inert such as ground clay, talc, or chalk, which functions as a carrier. These products pose an inhalation hazard, but the inert ingredients are usually less hazardous. The inert of greatest concern here is usually silica. Granular and pellet formulations typically contain lower amounts of active ingredient bound to a larger particulate inert carrier such as ground vegetable material (e.g., corn cob, nut shell), sand, or clay.

5.7 Evaluation and Referral for Site Inspection or Enforcement

The enforcement agency that receives inspection referrals varies across States and circumstances. Most often it is the DA; less often the State occupational health agency, department of forestry, or department of environmental protection. In most instances, the PPSP cannot conduct an investigation or evaluation without having some contact with enforcement agencies. Each PPSP must make its own decisions about referral protocols for investigation by these agencies. The PPSP should consider several issues including confidentiality concerns, whether other persons are at risk, whether the exposure is ongoing, the severity of the illness or injury, and whether circumstances suggest possible rule violations. There should be a system for documenting and tracking referrals to determine the outcome of investigations by the partner agencies, and to ensure that findings are disseminated to the appropriate persons and agencies. It is helpful to have written protocols that describe the priority system and process for referral.

When the PPSP makes a referral, a person's name is not usually released unless the person has agreed to the release or agrees to contact the enforcement agency him- or herself. Case intake forms should have a place to indicate if verbal permission has been given to release sufficient information for a referral, and whether this includes permission to release the person's name. PPSP must develop guidelines on when a written release is required. This issue is usually of greatest concern for occupational exposures where workers are concerned about loss of employment, and nonoccupational exposures where tenants are concerned about loss of housing.

The PPSP must decide under what circumstances a referral is made to an enforcement agency without the permission of the affected person. Exposed workers may be reluctant to have the PPSP contact their employer or make a referral to an enforcement agency. When deciding how to proceed in these situations, the PPSP must take into account whether others are at risk, the nature of the exposure, and the severity of the illness or injury. Affected persons should be informed both of their rights about confidentiality, and that some enforcement agencies may be less able to protect confidential information compared with the PPSP.

5.8 Overview of Agencies with Jurisdiction of Pesticides

This section outlines the principal agencies typically involved in regulating pesticide manufacturing, distribution, use, and disposal at the State level. As the many entries suggest, a great deal of variability exists in how individual States manage oversight of pesticide use. There is no uniform way that Federal rules are incorporated into State laws/rules, nor is there consistency in the State agencies designated to implement or enforce these laws/rules. (Appendix F provides an overview of the main Federal rules that relate to pesticide use and potential exposures.)

5.8.1 FIFRA State Designees

The key agency for enforcing most of the rules mandated by FIFRA varies from State to State. In most States, the DA functions as the EPA designee for enforcement of FIFRA. In others, the department of environmental conservation, or other similar agency holds that responsibility. The areas covered by FIFRA include the registration of pesticide products, product labeling, licensing of pesticide applicators, the sale and distribution of pesticides, and proper work practices for handling pesticides. Rules on the disposal of pesticides are usually enforced by the State environmental agency, which will also be involved in responding to spill events or events involving pesticide contamination of bodies of water.

The part of FIFRA governing proper agricultural work practices is WPS, the provisions of which are usually enforced by the same State agency that enforces FIFRA. Formal agreements designate the primary State agency responsible for WPS enforcement. Oregon is currently the only State to delegate WPS enforcement to a State OSHA program, by agreement with the Oregon DA. In Washington State, both the DA and the Department of Labor and Industries simultaneously adopted identical WPS rules based on EPA standards. A detailed description of FIFRA is included in Appendix F.

5.8.2 OSHA

Twenty-six States/territories operate their own OSHA program. In most of them, the OSHA program is a part of the DOL, but in others it may be part of the insurance division, health department, or other section of the State government. The groups of workers covered varies among the different State and territorial programs. For more information about State programs, see http://www.osha.gov/fso/osp/index.html. In States without their own OSHA program, the Federal OSHA is responsible for regulating occupational safety and health. Federal OSHA maintains one or more regional offices in these other States.

The Federal Occupational Safety and Health Act contains provisions that relate to pesticide exposure, including hazard communication, farm labor housing, field sanitation, agriculture, fumigants, first aid and emergency services, and general duty clauses about provision of a safe and healthy workplace. OSHA also has responsibility for workers involved in manufacturing and formulating pesticides.

State level OSHA programs may have broader jurisdiction over occupational health issues related to pesticides. PPSP staff should seek training and information from OSHA staff in their State about State rules and agency jurisdiction in this area.

5.8.3 AGENCIES RESPONSIBLE FOR DISEASE SURVEILLANCE AND CONTROL

Responsibility for surveillance of both environmental and occupational pesticide-related illness and injury may be in the same office within an agency, or they may be scattered in different State agencies and offices. The simplest situation is one in which they occupy a single office within an agency. This allows development of protocols that encompass occupational and nonoccupational exposures. In addition, single exposure events that involve occupational and nonoccupational cases can be managed by the same staff. If the two types of cases (that is, occupational and nonoccupational) are managed in different offices within the same agency, a central point of contact for all cases is usually easiest for reporters. This central contact should have well-structured referral and follow-up procedures.

Many States participate in a centralized system established for reporting chemical spills and releases, called the Hazardous Substances Emergency Events Surveillance (HSEES) system. Funding to develop and maintain HSEES is provided by the Agency for Toxic Substances and Disease Registry (ATSDR). The PPSP should make sure it is linked to any HSEES activities in its State, and receives reports of events involving human exposures to pesticides.

5.8.4 OTHER STATE AND ADJUNCT AGENCIES

Other agencies that may be involved in case investigations include the State agencies responsible for the following:

- Environmental regulation (that is, the PPSP is usually involved with these agencies when an event includes issues related to disposal, transport, spills, or other significant environmental contamination, or releases into bodies of water, air, or soil),

- Forestry (that is, when an event involves applications to State forest lands), and

- Fish and wildlife (that is, when an event includes harm to fish or wildlife).

5.8.4.1 Vector Control Districts

Vector control districts are responsible for control of disease vectors at the county or regional level. They provide public education to help control breeding of rodents, mosquitoes, flies, and ticks. They also conduct pesticide applications to control disease vectors and nuisance problems caused by the vectors. It is helpful to have a list of the local districts and information about what pesticide products they are using. This will allow the PPSP to contact the appropriate district upon receipt of illness reports that are associated with vector control activities. The American Mosquito Control Association's Web site (www.mosquito.org) has links to affiliated mosquito and vector control associations.

5.8.4.2 Pest Control Boards

Some States have governing bodies involved in the regulation of nonagricultural commercial pesticide applications, such as applications by structural pest control operators. These bodies can be helpful in case reporting, investigating, and developing intervention strategies.

5.9 State Interagency Coordination of Case Investigations

Outlining the most effective structure for interagency coordination is extremely difficult because the level of cooperation, available resources, and expertise across State agencies is not standard. This section discusses several of the different approaches used to address interagency coordination. At a minimum, it is important to know which agencies in the State have responsibility for the various issues associated with pesticide incidents and to have a list of the appropriate contacts in each of those agencies.

5.9.1 Interagency Agreements

Washington and Texas PPSPs each maintain a memorandum of understanding between State agencies for investigation of pesticide poisoning cases. The agreements set forth formal arrangements among State agencies about communication, responsibilities, and jurisdiction for investigation of pesticide-related health complaints. Formal interagency agreements can be time-consuming to negotiate and may end up with rigid clauses that do not provide sufficient flexibility to address all situations that may occur. Nevertheless, they are helpful in clearly stating roles and responsibilities of agencies and setting a clear structure for cooperation. The existence of formal agreements also sets a precedent for documented cooperation that is easier to maintain over time as agency management and personnel change.

5.9.2 Multiagency Coordinating Boards

Two States (Oregon and Washington) have created multiagency boards to establish mechanisms for coordinating investigations, evaluating data from investigations, and developing action plans for pesticide poisoning prevention. These boards are designed specifically to address adverse human and environmental impacts from pesticide use and are briefly described below. Web site addresses for the statutory language establishing the two programs are provided in Appendix B. States interested in pursuing development of similar boards are advised to contact the Oregon Pesticide Analytical and Response Center (PARC) and Washington Pesticide Incident Reporting and Tracking Review Panel (PIRT) to obtain annual reports and other information about these programs.

PARC was established in the early 1970s in response to public concerns about the health effects from herbicide spraying conducted by the forestry industry. The board is composed of representatives from seven State agencies with jurisdiction over pesticides or health, a representative from the Oregon Poison Center, and one citizen appointed by the governor. Various toxicologists within the State university system are included as consultants to the board. PARC is designed to centralize reporting of actual or alleged health and environmental incidents involving pesticides. It is also designed to mobilize the expertise needed to investigate pesticide incidents in a timely manner. The board examines data to identify trends and problems, and may make recommendations for actions to member agencies. The budget for PARC was eliminated in 2003, but agency members continue to meet on a regular basis to discuss investigations and review cases. The Washington PIRT panel was modeled on the PARC board but has somewhat broader mandates.

5.9.3 Advisory Committees

If a PPSP does not develop formal interagency agreements and there is no statutorily mandated multiagency oversight committee (or board) to address pesticide use, the program might benefit from developing an advisory committee. Members of the advisory committee can include representatives from other partner agencies, public interest groups (e.g., environmental and public health organizations), agricultural employers, worker advocacy groups, PCCs, HCP associations, pesticide manufacturers or reformulators with facilities in the State, and the pest control industry. The committee should meet two to four times per year. The meetings are often a source of valuable ideas to the program. They also provide the PPSP an opportunity to maintain contact with various constituencies, apprize them of findings, develop joint programs for outreach and intervention, and discuss mechanisms for improving reporting and investigation.

5.10 NIOSH, NCEH, AND EPA

Prompt notification of NIOSH, NCEH (for nonoccupational cases only), and EPA may allow those agencies to work with the State PPSP to prevent additional cases. Guidelines used to trigger NIOSH/NCEH/EPA reporting are case reports that involve any of the following:

- hospitalization or death from unintentional pesticide exposure, or

- events that involve 4 or more ill persons, or

- events that occur despite use according to the pesticide label, or

- events that indicate the presence of a recurrent problem at a particular workplace and/or with a particular employer's worksites.

Prompt sharing of information with Federal partner agencies alerts them to possible emerging problems and may trigger additional investigative action and assistance to the PPSP. NIOSH may be notified by contacting the Surveillance Branch, Division of Surveillance, Hazard Evaluations and Field Studies at 1-800-356-4674. NCEH may be notified by contacting the Health Studies Branch at 404-498-1340. Notification of EPA can include contacting both the EPA Regional Office (see Appendix G) and the Health Effects Division of EPA in Washington, D.C. (703-305-7576 or 703-305-5336). Ideally, notification between the EPA Regional Office and the PPSP should be reciprocal (that is, the PPSP should be notified by EPA when EPA learns of events within the PPSP's area of jurisdiction). The EPA Regional Office typically will refer complaints or reports to the State designee (that is, the State agriculture department). Establishing routine contact between programs will make reciprocal notification more likely. NIOSH and EPA may also assist in mobilizing the resources of other agencies as needed (e.g., to investigate the illegal residential use of methyl parathion in several States, the EPA regional offices solicited assistance from NCEH and ATSDR [EPA 1996]).

5.11 Federal Agencies That May Have a Role or Be a Resource During Case Investigation

This section describes those Federal agencies with which the PPSP will likely have the greatest contact.

5.11.1 United States Department of Agriculture (USDA)

Several USDA programs can serve as useful information resources or as partners in educational programs. The primary programs pertinent to pesticide illness surveillance are as follows:

- Cooperative State Research Education and Extension System
- Federal Grain Inspection Service
- Animal Plant Health Inspection Service (APHIS)

5.11.1.1 Cooperative State Research Education and Extension System (CSREES)

CSREES is a national system based in the land grant universities and county administrative units. This system is well recognized in rural communities, and increasingly in urban areas, as a source of information and practical classes. The system maintains agricultural experiment stations that work with university researchers, including toxicologists. Programs include (1) IPM, (2) sustainable agriculture, (3) food safety, (5) family health, (6) 4-H clubs, and (7) environmental and water quality programs. The EXTOXNET system is a resource on pesticide and environmental toxicology sponsored through CSREES. Local extension agents provide information about crops, seasonal pesticide use, and particular pest problems in local areas. They can be valuable resources in understanding local agricultural issues. The extension agents usually specialize in particular crops and may be a valuable resource in determining what products are typically used and can identify the local aerial applicators. The extension service works with EPA and State designees to conduct pesticide applicator training programs. The extension service also offers programs aimed at youth (e.g., 4-H), farm families, and suburban gardeners. These established training programs can be ideal avenues to disseminate pesticide safety information from the PPSP. The extension programs also can be ideal partners for developing and testing interventions. (More information about extension programs appears in Appendix G.)

5.11.1.2 Federal Grain Inspection Service (FGIS)

FGIS is part of USDA Grain Inspection, Packers, and Stockyards Administration. It establishes the methods and standards used to describe grain quality. FGIS or delegated State agencies conduct mandatory export grain inspections and other nonmandatory programs for domestic grain commerce. PPSPs will usually not have much contact with this branch of USDA unless addressing fumigant exposures to FGIS grain inspectors, or to the public from treated grain or vehicles (railcars, barges, etc., used to transport grain).

5.11.1.3 Animal Plant Health Inspection Service (APHIS)

APHIS is responsible for conducting activities aimed at protecting agriculture in the United States. These activities include securing the U.S. borders against foreign agricultural pests and diseases, as well as facilitating exports of agricultural products. It is also involved with preventing damage to agriculture from wildlife (including through the use of pesticides). Additionally, APHIS is involved in ensuring the safety of genetically engineered plants and other agricultural biotechnology products. At the State level, APHIS works with State departments of agriculture and health when planning emergency actions associated with elimination of foreign pests. Recent examples of these types of activities include programs to eradicate medfly and citrus canker in Florida, medfly in California, and Asian or European Gypsy Moth in many States. It is important that PPSP programs work closely with DA and APHIS contacts on these types of eradication programs. These eradication programs may require significant levels of public education and outreach activities aimed at HCPs and the public prior to a pest control operation taking place. It is extremely important for the various agencies involved in emergency actions to present the same risk communication message since differing messages can weaken public trust and understanding. If the infestation is a regional problem, activities may be coordinated with other States in the region.

When addressing these eradication programs, the PPSP may choose to add a more active component to its routine passive surveillance. There also may be reasons for conducting a more structured epidemiologic study to address particular concerns. This might include controlled studies of applicators, or monitoring emergency room reports for particular illnesses of concern and comparing illness rates with background levels [Green et al. 1990; Pearce et al. 2002]. The volume of calls can increase significantly surrounding these types of spray programs, resulting in a considerable increase in workload. During these events, PPSPs frequently set up hotline operations to deal with complaints and questions. In addition, agency Web sites can serve as valuable sources of information for the public. Fact sheets and up-to-date spray schedules and maps can be posted and updated as frequently as needed in response to changing conditions (for an example, see the New York State Department of Health Web page on West Nile Virus at http://www.health.state.ny.us/nysdoh/westnile/index.htm).

5.11.2 Federal Aviation Administration (FAA), National Transportation Safety Board

The agency that investigates airplane accidents can provide information about airplane accidents involving aerial pesticide applications. Reports on investigations are available on the agency Web site (see Appendix C for the address and instructions on conducting a search).

5.11.3 U.S. Fish and Wildlife

This agency has responsibilities for protection of wildlife and may be involved in investigations of wildlife poisoning. They and their State partner agencies may work with the PPSP on analyses to determine whether pesticides are implicated in wildlife deaths that may also involve potential human exposures. The agency is also active in issues associated with pesticide use and the Endangered Species Act.

5.11.4 OTHER FEDERAL AGENCIES

A number of other Federal agencies, along with their State counterparts, can be helpful during some investigations. These agencies will be discussed briefly because PPSPs will only periodically collaborate with them.

5.11.4.1 COAST GUARD

The Coast Guard will play a role in addressing exposure incidents involving spills in navigable waterways.

5.11.4.2 CONSUMER PRODUCT SAFETY COMMISSION (CPSC)

The CPSC may be a useful partner when addressing issues associated with imported products, such as insecticidal chalk.

5.11.4.3 CUSTOMS BUREAU

The Customs Bureau should be notified when information is obtained about importation of illegal pesticides (that is, pesticides not registered for use in the United States).

5.11.4.4 DEPARTMENT OF TRANSPORTATION (DOT)

DOT plays a role in the regulation of interstate shipping of pesticides and may be a useful contact for exposure incidents involving a shipping accident or spill.

5.11.4.5 FEDERAL BUREAU OF INVESTIGATION (FBI)

The FBI should be consulted when PPSP suspects malicious use of pesticides, and the malicious use has potential community or broad public impact.

5.11.4.6 FEDERAL RAILWAY ADMINISTRATION

The Federal Railway Administration may be involved in situations involving rolling stock (that is, rail cars anywhere other than in rail yards or depots that are under OSHA jurisdiction). They will also have a role in addressing releases of pesticides being transported by rail.

5.11.4.7 FOOD AND DRUG ADMINISTRATION (FDA)

The FDA may be involved with investigations involving veterinary or pharmaceutical uses of pesticides, and genetically modified crops with pesticidal properties.

6. Site Inspections by PPSP

6. Site Inspections by PPSP

6.1 Introduction

Most poisoning events can be investigated simply through a telephone interview with the poisoned subject, combined with additional information gleaned from medical records and investigation reports from enforcement agencies. In some instances, however, a site inspection by the PPSP may be necessary. The process for initiating an inspection depends on the agency's authority to access the type of site involved. Section 2.5.1.9 *Authority to Investigate* discusses this in more detail. It's worth remembering that site inspections are very resource intensive.

In most States with established PPSPs, investigations are conducted by program staff based in the centralized State office. The geographic location of these surveillance programs can present a significant drawback when conducting site inspections, since travel can take a significant amount of time. In contrast, the Washington State program has sufficient staff to base investigators in several areas of the State.

Each PPSP must set its own criteria for what triggers a site inspection, bearing in mind that specific mandates may need to be followed. These mandates may be established by a funding source, the demands of special projects, or general requirements of the disease reporting rules in the State. Criteria used to trigger a site inspection may include the following:

- All deaths from nonintentional exposure
- All hospitalizations from nonintentional exposure
- Four or more ill persons associated with a single exposure event
- An unusual temporal clustering of three or more reports associated with a particular pesticide product (especially those newly on the market), ingestion of pesticide-treated food, a pesticide device, or a particular workplace/employer
- Incidents involving a pesticide, class of pesticide, type of application, or industry selected by the surveillance program for special emphasis

Obtaining the cooperation of the affected persons and the employer or owner of the exposure site is critical if an inspection is conducted. The surveillance program must develop protocols covering whether employers will receive advance notice prior to inspections. This decision is usually based on whether the agency has a clearly mandated enforcement responsibility for any facet of the State's implementation of the Occupational Safety and Health Act (OSH Act) or FIFRA. If the surveillance program functions outside those acts, the program investigators may decide to offer inspections as a form of free consultative service to the employer, business owner, or home owner. In these cases, an advance phone call may help establish the foundation for a cooperative relationship. If the primary purpose of the inspection is to obtain a completely unbiased view of the operation, it may be more useful to perform unannounced site visits. This may not be feasible, however, unless conducted jointly with an enforcement agency.

If the program does not have a formalized authority to investigate, a pre-established plan of action should be in place to handle inspection refusals. The carrot-and-stick approach is often effective. The agency can discuss the benefits of allowing an inspection on a cooperative basis, namely that the investigator will provide information and assistance in preventing exposures. The owner or employer may avoid a formal referral to an enforcement agency by agreeing to correct the hazards identified during the inspection. Care should be taken to ensure that the employer understands the voluntary nature of the inspection, and what actions will be taken if an imminent danger situation is identified. Other items that should be explained include the following:

- The scope of the inspection

- An explanation of what information will be held confidential, if any

- What types of actions will be taken when problems are identified

- Inability of the inspection to identify all hazards or violations of good practice

- An explanation that cooperating with the inspection and/or following recommendations made in the inspection report will not exempt the employer or worksite from an enforcement inspection or complying with relevant regulations

- The information that will be provided at the end of the inspection and to whom it will be provided

It is a good idea to develop standard language covering these elements and to provide it in writing both at the beginning of the inspection and in the final report.

In some situations, PPSP staff may choose to conduct site inspections simultaneously with the enforcement agency, depending on the relationship between agencies. Employers can feel besieged when multiple agencies conduct separate inspections at different times since these activities disrupt normal work activities. Alternatively, the PPSP may prefer to keep its inspections separate to prevent confusion with another agency's mandates, or to maintain a different relationship with the owner of the establishment and the exposed persons. In these cases, the investigator must be able to explain why the PPSP inspection is different, what is being evaluated, and what type of information will be provided to the employer and employees at its conclusion. (Site inspections of nonoccupational exposures involve similar issues with landlords, public buildings, neighboring property owners, and retail establishments.)

PPSP staff may be contacted during emergency response events, such as spills or fires involving pesticides. Programs should have policies to address staff roles in these circumstances. Any on-site work during these events requires that staff have the proper level of safety and health training and PPE.

6.2 Getting Started with the Inspection

Attire for site inspections should be appropriate for the type of establishment. Failure to dress accordingly will hinder the investigator's ability to establish a credible working relationship with all of the persons involved in an exposure event.

The investigator should begin by introducing him- or herself by name, title, and organization, and presenting appropriate credentials. The purpose for the visit should be provided next. The investigator should meet with company and worker representatives to discuss the timetable and purpose of the visit and to obtain information about the exposure event.

6.3 Site Walk-Around Evaluation

The purpose of the walk-around inspection is to gather information to

- Evaluate the relationship between the reported illness and the pesticide exposure

- Identify potential safety and health hazards related to pesticide use in the home or workplace

- Document the exposure

- Observe the activities of affected and other potentially exposed persons

- Identify changes in policies or procedures that will help prevent the recurrence of a similar exposure event.

It is useful to diagram the site where the exposure occurred and indicate the location of any windows and ventilation ducts. Another helpful step is to review relevant written policies, training program materials and records, and the injury and illness log. Finally, investigators should obtain multiple perspectives for why the exposure event occurred. In occupational exposures, that includes the exposed worker(s) and either their employer or supervisor.

Some programs conduct only limited worker interviews during a site inspection. Others conduct those interviews only at the worker's home, by telephone, or at a neutral place outside of work hours. In most situations, in-depth worker interviews are most effective when conducted away from the workplace. At the workplace, time spent talking with investigators may compromise the confidentiality of the exposed worker, decrease the worker's earnings (especially in agricultural settings), or make it difficult for the worker to provide as much information as might be possible away from work.

6.4 Equipment for Site Inspections

6.4.1 Camera

A camera is indispensable for quick documentation of the site layout, but photos need to be augmented by notes, diagrams, and measurements. Photographs can help document the state of repair of application equipment, PPE, or pesticide product storage. Photographs are also useful in documenting sampling sites. (Sampling equipment is described in Section 6.5.) When photographing workers, consider how the photographs will be used. Formal consent should be obtained if the worker will be identifiable. As a courtesy, explain to workers how the photographs will be used even if the workers will not be identifiable.

6.4.2 PPE

The information provided here is general and is designed to serve as a reminder to State programs that they need to address issues of staff safety and health. It is extremely important that staff conducting inspections are equipped with appropriate PPE for the types of situations they will be evaluating. If there is any doubt about the safety of entering a particular area with the level of PPE available, the inspection should be terminated until appropriate PPE is obtained. All staff evaluating pesticide illnesses must be in compliance with all safety and health rules for their own protection and to protect the credibility of the program. Guidelines for appropriate PPE should be reviewed by the agency's safety and health officer or other appropriate staff to ensure compliance with occupational safety and health laws. Staff should be properly fit-tested for respiratory protection devices. (Guidelines for fit-testing and medical evaluation for use of respiratory protection are included in the OSHA Respiratory Protection Standard [29 CFR 1910.134]).

The minimum equipment that should be available for inspections is as follows:

- Chemical resistant boots (polyvinyl chloride [PVC], nitrile, or similar material)
- Half-face cartridge respirator or powered air purifying respirator (PAPR) equipped with cartridges appropriate for the hazard (usually a combination organic vapor cartridge with a dust filter)
- Unlined gloves of nitrile or butyl rubber
- Steel-toed rubber boots
- Plastic bags for transporting PPE that may be contaminated during an inspection
- Any necessary sampling equipment and containers for transporting samples

6.4.3 Water

Staff should carry sufficient water to ensure proper hydration if working in a hot environment. Water should also be available for decontamination when staff members are observing pesticide mixing, loading, or application activities.

6.4.4 Contact form

Since inspections are frequently conducted in remote areas, it is helpful to have staff complete a simple contact form prior to leaving the office. The form should identify the location of the inspection site, the staff departure time, the expected return date and time, and a checklist of PPE to be taken to the inspection site (see Appendix C).

6.5 Sampling

This manual provides general information about sampling. Sources for more detailed instructions on sampling and laboratory pesticide analyses are described in this section. Several types of sampling are appropriate for ascertaining whether exposure occurred or was probable, whether pesticide was absorbed, and whether changes in biologic function took place as a result of exposure. In most situations, PPSPs do not have adequate funding and staff to conduct routine environmental or biological sampling. Sampling is most often conducted as part of enforcement inspections by agencies that have jurisdiction over pesticides or occupational health. The sampling carried out by these agencies is typically aimed at determining whether a code violation occurred. This may be different from the sampling desired by a public health agency aimed at ascertaining whether a person was exposed to sufficient pesticide to suffer health problems, or whether use of a product according to the label may cause adverse health effects.

State enforcement agencies may maintain sampling guides or manuals. PPSPs should review these if they plan to conduct their own site sampling or use sampling results from the State enforcement program. NIOSH, OSHA, and EPA maintain manuals of analytic methods for a broad range of chemicals including pesticides [NIOSH 1994; OSHA 2000; EPA 2000a]. Chapter 13 and Appendix A of the FIFRA Inspection Manual [EPA 2002] also contain procedures for collecting residue and environmental samples.

Pesticide manufacturers can be a useful source of information about sampling methodologies for their products. Their industrial hygienists can provide useful information about sampling methods and data from the company's exposure analyses. They are also a good source for information about decontamination procedures following significant spills or misapplications of their products.

With the exception of equipment for surface wipe sampling, most sampling protocols require an investment in equipment for sampling and calibration. It is possible to rent or develop a system for borrowing equipment that will not be used frequently. If sampling is not conducted regularly by experienced staff, sampling measurements may be inaccurate due to various sampling errors. PPSPs that do not conduct regular sampling may choose to arrange for it to be conducted by a sister agency (e.g., an enforcement agency) through an interagency agreement.

6.5.1 Sample Collection

The sample collection strategy is dictated by the purposes for obtaining the samples and the circumstances of the exposure. The date, time, and environmental conditions of sampling must be carefully recorded. Storage and handling of the sample must be documented.

Residue samples are typically the most common types of samples encountered in health inspections. These involve obtaining samples of plant material, animal tissues, water, soil, wipes of hard surfaces, or samples of contaminated fabric, air, runoff water, etc. At times, samples may be taken to determine whether the tank mix and/or dilution of pesticide was appropriate according to the product label or to identify an unknown pesticide product.

In the case of drift exposures, it is useful to take residue samples on the actual site where the application was intended. The area from the site of the application to the site where affected persons were exposed should be divided into grids. A series of samples is then taken in each grid, moving from the site of application to the site of exposure. Samples of soil or foliage are most commonly used to document drift. Wipe samples taken from vehicles or building structures may also be useful for documenting drift exposures. Contaminated clothing may be collected, although analysis is often more difficult.

Sampling pumps must be carefully calibrated before and after any air sampling. Indoor air samples must address issues of potential interference from other indoor air contaminants. In residential exposure situations, samples should be taken under the same conditions that existed at the time the person was exposed (e.g., with respect to heat and ventilation). In addition, samples should be taken under a worst-case scenario with heat on or air conditioning off to determine any ongoing hazard from exposure in the residence.

If biological specimens such as blood or urine are collected, it is critical that the analytical laboratory be contacted ahead of time. The laboratories' instructions about sample collection media, preservatives, storage conditions for transport, and shipping must be carefully followed.

6.5.2 Sample Preparation Custody and Handling

The investigator should carry clean sampling materials, container seals, and preservatives as needed. Proper care of the sample during transport is critical for sample integrity. Chain of custody should be documented using a standard form. The manuals described earlier have examples, as do State enforcement agencies. The laboratory that will be receiving the sample may require a chain of custody form.

7. Case Closure and Classification

7. Case Closure and Classification

7.1 The Case Closure Process

Once interviews have been completed, medical information obtained, and any sampling or site inspection conducted, the investigator should evaluate the case for completeness. It may also be helpful to have each case reviewed for completeness by an additional investigator. A list of any additional information needed should then be developed and additional follow-up conducted. After all pertinent information has been collected, the report should go through a formal evaluation and classification. There is often an extremely long delay in obtaining reports from enforcement agencies, so the PPSP may choose to develop a provisional case closure protocol, with final case closure occurring after receipt of the enforcement investigation report.

Case classification is often performed as a group process to help ensure objective and consistent evaluation of all cases within the surveillance program. Some PPSPs have a panel of persons who individually review and classify each case. Afterwards, all panel members' classifications are reviewed, and the panel meets to discuss any differences. Other programs assign two persons who classify cases as a team, with a designated third party to consult when the pair disagree on the interpretation of available case data.

Case information should be reviewed using the matrix in the *Case Definition for Acute Pesticide-Related Illness and Injury Cases Reportable to the National Public Health Surveillance System* [NIOSH 2004]. The reviewers may sometimes feel there are circumstances with a case that should result in a classification that is different from that obtained by using the matrix. They should record the classification they think is correct and note the reasons for any differences with the matrix classification. If the PPSP is using a separate State classification system in addition to the national case definition, this classification should also be applied at the time of case closure.

7.2 Case Definition for Acute Pesticide-Related Illness and Injury Cases Reportable to the NPHSS

The case definition described here was developed by a consensus process involving Federal and State agency partners and was adopted by CSTE [1999]. The full text of the case definition and its appendices are included in Appendix D. Portions are provided here for discussion. The case definition provides a consistent, objective approach for assessing information gained about each report. Using this definition to evaluate and classify cases allows an aggregation of data from all States and a comparison of data between States.

7.3 Clinical Description

7.3.1 Case Definition

This surveillance case definition refers to any acute adverse health effect resulting from exposure to a pesticide product (defined under

FIFRA, except that disinfectants are often excluded[§]) including health effects due to an unpleasant odor, injury from explosion of a product, inhalation of smoke from a burning product, and allergic reaction. Because public health agencies seek to identify and prevent all adverse effects from regulated pesticides, notification is needed even when the responsible ingredient is not the active ingredient.

A case is characterized by an acute onset of symptoms that are dependent on the formulation of the pesticide product and involve one or more of the following:

- Systemic signs or symptoms (including respiratory, gastrointestinal, allergic, and neurological signs/symptoms)

- Dermatologic lesions

- Ocular lesion

This case definition and classification system is designed to be flexible in permitting classification of pesticide-related illnesses from all classes of pesticides. Consensus case definitions for classes of chemicals may be developed in the future.

A case will be classified as occupational if the person is exposed while at work (this includes working for compensation; working in a family business, including a family farm; working for pay at home; and working as a volunteer emergency medical technician [EMT], firefighter, or law enforcement officer). All other cases will be classified as nonoccupational. All cases involving suicide or attempted suicide should be classified as nonoccupational.

A case is reportable to the national surveillance system when the following criteria are met:

- Documentation of two or more new adverse health effects that are temporally related to a documented pesticide exposure, combined with

    a. consistent evidence of a causal relationship between the pesticide and the health effects based on the known toxicology of the pesticide from commonly available toxicology texts, government publications, information supplied by the manufacturer, or two or more case series or positive epidemiologic investigations, *or*

    b. insufficient toxicologic information is available to determine whether a casual relationship exists between the pesticide exposure and the health effects.

7.3.2 Laboratory Criteria for Diagnosis

If available, the following laboratory data may confirm the diagnosis of acute pesticide-related illness or injury:

- Biological tests for the presence of, or toxic response to the pesticide and/or its metabolite (in blood, urine, etc.)

---

§ Pesticides are defined under the Federal Insecticide Fungicide and Rodenticide Act (FIFRA) as any substance or mixture of substances intended to prevent, destroy, repel, or mitigate insects, rodents, nematodes, fungi, weeds, microorganisms, or any other form of life declared to be a pest by the Administrator of the U.S. EPA and any substance or mixture of substances intended for use as a plant regulator, defoliant, or desiccant. Pesticides include herbicides, insecticides, rodenticides, fungicides, disinfectants, wood treatment products, growth regulators, insect repellents, etc.

Please note that adverse health effects resulting from exposure to disinfectant products are not reportable in many States because the volume of reports could overwhelm the State's surveillance system; therefore, these cases will not be routinely reported to the national surveillance system. Certain States may collect data on health effects resulting from a few selected disinfectants (e.g., glutaraldehyde). If resources are available, States are encouraged to do surveillance on acute disinfectant-related illness.

- Measurement of the pesticide and/or its metabolite(s) in the biological specimen

- Measurement of a biochemical response to the pesticide in a biological specimen (e.g., cholinesterase levels)

■ Environmental tests for the pesticide (e.g., foliage residue, analysis of suspect liquid)

■ Pesticide detection on clothing or equipment used by the case subject

7.3.3 Case Classification Categories

The case classification matrix is used to rank the evidence linking the illness/injury to the pesticide exposure. Only reports meeting case classifications of Definite, Probable, Possible, and Suspicious are reportable to NPHSS. Additional classification categories are provided for States that choose to track reports that do not fit the national reporting criteria.

Classification categories:

Definite: Objective evidence confirms the exposure and illness, and the temporally related illness is consistent with the known toxicology of the pesticide.

Probable: Objective evidence of either the pesticide exposure or the health effects is available, and the temporally related illness is consistent with the known toxicology of the pesticide.

Possible: Only subjective evidence of exposure and illness is available, and the temporally related symptoms are consistent with the known toxicology of the pesticide.

Suspicious: Insufficient toxicological information is available to determine whether a causal relationship exists between the pesticide exposure and the health effects.

Unlikely: The relationship between the reported exposure and illness is not consistent with the known toxicology of the pesticide.

Insufficient Information: Insufficient documentation was obtained about the exposure or health effects to determine whether the health effects were related to a pesticide exposure.

Not a case: A person was reported to a State surveillance system due to an alleged exposure, but was asymptomatic, or it was determined that health effects were due to a condition other than a pesticide exposure. States may choose to create a subset of the *not a case* category to track asymptomatic persons exposed to pesticides.

Once a case has been classified, a determination of illness/injury severity can be performed using the severity index provided in Appendix E. Severity is assigned only to cases reported to NPHSS.

7.4 Communication of Findings and Recommendations

PPSPs should aggregate individual case data to describe the magnitude and distribution of pesticide illness and injury, develop and target prevention messages, and develop policy. However, more immediate feedback to reporters (that is, to HCPs and other sources of case reports) is critical for maintaining reporting to the surveillance program and ensuring that this very teachable moment is used to deliver prevention information. Some programs send a routine letter to providers to thank them for reporting. In the case of nonreporting providers, the PPSP can send notification to the provider indicating that the PPSP is investigating a poisoning report involving the provider's patient. Several PPSPs send a letter to providers when a case is closed, providing a brief summary of findings and the epidemiologic case classification. Another useful gesture is to provide a copy of *Recognition and Management of Pesticide Poisonings* [Reigart and Roberts 1999].

All site inspections should result in a written report of findings provided to the affected person(s), and the employer or third party responsible for the pesticide application. In the case of worksite inspections, a brief summary report may be provided to all interviewed workers. This report should clearly communicate any recommendations, including those for IPM, product substitution, use of PPE, and training or engineering controls. In some situations, worker representatives (e.g., union or advocacy organizations) are involved. These worker representatives should be given a copy of the written report, as they can serve as another venue for making sure workers receive and understand inspection findings. Care should be taken to make sure confidential information is excluded from reports sent to affected persons, workers, employers, and other third parties. Follow-up to determine whether recommendations are adopted should be conducted after an appropriate time interval. Follow-up may be conducted by mail, telephone, or a site visit, especially if recommendations included engineering controls.

8. Data Analysis and Reporting of Aggregated Data

8. Data Analysis and Reporting of Aggregated Data

8.1 Introduction

Data analysis is an important component of surveillance. Data analysis is required to decipher the message that the surveillance data are trying to provide. As such, it will assist in identifying the most problematic pesticides, the risk factors associated with acute pesticide-related illness, and emerging data trends. This chapter will briefly discuss analysis of pesticide poisoning surveillance data. A more detailed review appears in standard epidemiological and surveillance textbooks [Teutsch and Churchill 2000; Rothman and Greenland 1998].

Data dissemination is another important function of PPSPs. The control and prevention of acute pesticide-related illness depends on the dissemination of surveillance data to ensure that educational, consultative, and regulatory interventions can be effectively targeted. In addition, surveillance partners (that is, HCPs, other government agencies, and labor and industry groups) welcome reports on surveillance findings and their impact. Keeping partners informed can promote visibility and support for PPSP. Additional details on the content and audience for PPSP surveillance reports are provided in this chapter.

8.2 Routine Descriptive Analysis

The pesticide poisoning data compiled by PPSPs can be analyzed in terms of person, place, and time. These analyses are useful for identifying variations that may be amenable to intervention.

8.2.1 Person-Based Analyses

8.2.1.1 Case Series

The case series focuses on persons and is the most basic level of data analysis and presentation. It serves as a useful tool for raising attention about emerging pesticide problems if they involve a particular pesticide or a particular type of pesticide usage (e.g., mosquito control or pesticide use in schools). A case series provides a narrative of the cases, including a description of the circumstances that led to exposures, and recommendations for preventing similar events. It also presents a distribution of case characteristics that can include disease classification, severity category, occupational versus nonoccupational, occupation, age, sex, and other characteristics that highlight the subgroups accounting for the greatest (and perhaps least) number or percent of cases. Reports of new and novel emerging problems that are reported in a timely manner should be considered for publication in the MMWR or another peer-reviewed publication. Examples of case series reports that were published in the MMWR appear in Appendix A.

8.2.1.2 Bivariate Analyses

Bivariate analyses involve examining the association between two variables. For example, an analysis of pesticide active ingredients and severity categories can assist in identifying the most problematic pesticides. Bivariate analyses can provide clues about risk factors that will allow the PPSP to target particular populations for outreach, education, or further study. Additional bivariate analyses can examine pesticide chemical class or functional class by risk factors such as demographic characteristics, occupation, industry, type of exposure (e.g., drift, spray, or contact with pesticide residue), or activity of the person at the time of exposure. The clustering of cases in particular worksites or across an industry can be used to determine whether a worksite or industrial group meets program criteria for targeted on-site investigations.

8.2.1.3 RATES

Ideally, risk factor data should be expressed as rates. The general form of a rate is:

$$\text{Rate} = \frac{\text{number of cases in specified time}}{\text{average or mid-interval population}} \times 10^n$$

where the numerator represents the number of cases meeting the specified criteria (e.g., males or farmworkers with acute pesticide-related illness), the denominator represents the size of the population during the specified time period (e.g., 1 year), and the size of n generally ranges from 2 to 6 depending on the size of the numerator (that is, n determines if the rate will be as low as *per 100 population* or as high as *per million population*).

The pesticide poisoning data collected by PPSPs is placed in the numerator. The denominator data must usually be sought elsewhere. When calculating county-wide or State-wide rates, an important source of denominator data for population estimates is the U.S. Census. When calculating rates among working populations, the denominator data can be obtained from the CPS, which is conducted by the Bureau of Labor Statistics (BLS) and is a useful source of employment population estimates.

Note that the denominator data involve limitations. Because the numerator and denominator data are often obtained from different sources, it is important to ensure that similar criteria are used for deriving the numerator and denominator. That is, the denominator should represent that portion of the population at risk for the exposure of interest. For example, when calculating rates of occupational pesticide poisoning, the denominator should consist of the number of employed workers, and not the total workforce or the total population, because the total workforce includes the unemployed who are not at risk of occupational pesticide poisoning. Similarly, when the denominator is the State population, the numerator should contain only residents of that State, because nonresidents who were poisoned in the State will not be included in the denominator. It is also important that the numerator and denominator are from the same time period (e.g., from the same calendar year).

Many factors influence the reliability of denominator (population) estimates, especially among agricultural workers. For example, migratory patterns and difficulties with tracking undocumented workers make it difficult to reliably count the total number of farmworkers. Another way of approaching the issue of denominators is to utilize the amount of pesticide used or sold in the geographic area in question. This is feasible only in States that routinely collect this type of data (e.g., California and New York). Even in these States, pesticide use information is very limited, as data collection is often confined to restricted-use products or products applied by licensed applicators.

Despite the limitations, the calculation of rates is very important. In contrast to simple counts, a rate allows a more informed comparison across groups (e.g., occupations, industries, age categories, sex, race, counties, States, etc.). Simple counts may erroneously suggest that a problem exists within a demographic group that accounts for a high proportion of the total population. By calculating rates, it may be found that the number poisoned per 100,000 persons does not vary across demographic groups. Examples of the use of rates appear in some SENSOR-pesticides publications [Calvert et al. 2003, 2004].

8.2.1.4 TIME-BASED ANALYSES

Data can also be analyzed to determine whether trends occur over time. Data comparing the number of cases reported in different years can indicate whether there were sudden shifts in the pattern of occurrence versus a stable number of reports. Although such data can be presented in tables, use of graphs can greatly facilitate the identification of trends and aberrations. Graphs can also provide visual clues on the impact of changes in case definition, program outreach efforts, introduction of a new pesticide product, or regulations restricting the use of a product. If the size of the population changes over time, it is more appropriate to examine time trends by graphing rates and not counts.

When constructing graphs, it is preferable to keep them simple. A simple, straightforward graph is much easier to interpret than a complex, cluttered graph. The graph should be able to stand alone. This requires having the title, axes, legend, and data source clearly and concisely labeled. Any data exclusions should be noted.

8.2.1.5 PLACE-BASED ANALYSES: MAPPING OF DATA

Mapping of data to show the geographic distribution of cases can be a useful tool for presenting information and for examining the spatial relationship between cases and sources of exposure. Many States with a PPSP provide a map indicating the number of cases by county. States currently involved in surveillance can geocode data and thereby produce detailed mapping using geographic information system (GIS) software. Geocoding involves including some way of recording the location of exposure that will permit geographic identification. This is not always feasible since descriptions of locations are not always clear (e.g., when workers do not know the address where they were working, pinpointing the exposure location is very difficult). The current version of SPIDER does not capture location data in a format needed for detailed geocoding (that is, location accuracy in SPIDER does not extend beyond address or Zip Code). It is anticipated that a future release of the SPIDER program will include fields (e.g., longitude and latitude) to record more precise exposure locations.

GIS mapping enables illness and injury data to be overlaid with information about crop distribution, location of waterways, roads, schools, etc. If the State has a system that records pesticide application data, this could also be mapped. Mapping can be useful to examine issues of large-scale vector control programs by mapping pest habitat areas, treatment areas, distribution of the disease of concern in animals or humans, and complaints and illness associated with pesticide use. Mapping can also be useful for identifying agricultural fields that abut residential, commercial, or school property, as these areas may be at risk for off-target pesticide drift.

This approach can be useful for generating hypotheses about risk factors that influence exposure, which can be further explored through epidemiologic studies. Data mapping may provide additional clues about geographic clustering of particular uses of pesticides or the prevalence of problems on the urban-rural interface. This information may be useful in developing interventions or providing information to policy makers responsible for land use or pesticide use regulations. The State of Washington is currently piloting a GIS approach for examining their data on agricultural pesticide-related illness and injury [Baum 2001b].

Any mapping of data must take into account confidentiality. A number of techniques are available to manipulate data to prevent identification of locations. Intentional errors (random shifts within a range) can be introduced into maps that contain street-level detail, making it difficult to determine exact locations. Maps with such introduced error must be annotated to make it clear that points are not exact locations. Simply presenting maps that do not provide street-level detail may also protect confidentiality. However, data presented on a regional level can still identify persons if they include age or race information. Other techniques to protect confidentiality include suppression for small numbers in detailed area maps or use of large symbols that interfere with the ability to precisely locate points [Spinello 2001].

8.2.2 Data Dissemination: Defining What Reports Will Be Useful For a Specific Program

Reports can be comprehensive or short and concise. Each PPSP must evaluate its own needs for different types of reports. It is important to determine the intended audience along with the purpose of the report, as the presentation of data will need to be constructed accordingly. A comprehensive summary report can provide an overview of the surveillance program, including tables and graphs on the overall magnitude of the problem, interpretation of the findings, and a description of recent program accomplishments. These reports are often required by Federal funding agencies and are also produced for multiagency coordinating boards. An annual report to legislatures must sometimes address statutory requirements (e.g., the need to demonstrate how the PPSP is meeting requirements for timeliness of response). There may also be a State interest in presenting data describing agricultural versus nonagricultural applications, illnesses affecting school students, or other issues. PPSPs with multiagency boards typically present data showing referrals from, or cases investigated by, partner agencies.

Reports prepared for the public or for HCPs should be brief–the content limited to one or a few issues. A case vignette or case series presented with some additional descriptive statistics may be useful for communicating a risk or prevention message to the public or engaging HCPs' interest in case identification, treatment, and reporting.

Reviewing reports produced by other PPSPs may provide other helpful ideas. In addition, the SPIDER program provides some standard descriptive tables that surveillance programs may find useful to include in their reports. See Appendix C for sample tables found in SPIDER. Finally, a recent article that provides national findings from the SENSOR-Pesticides Program may provide helpful ideas for report content [Calvert et al. 2004].

9. Developing Intervention Strategies and Evaluating Surveillance

9. Developing Intervention Strategies and Evaluating Surveillance

9.1 Designing Interventions Based on Surveillance Data

There are many types of interventions—education, adoption of engineering or administrative controls to mitigate exposure, implementation of more protective regulations, among others. The scope of these interventions may vary, with targets ranging from a single exposed person to the entire population, or from a single worksite to an entire industry. Interventions may involve a combination of techniques since adoption of even the simplest engineering control requires that some level of education and behavioral change also be adopted. Analysis of data to examine risk factors and identify areas for further investigation should be a routine activity of the surveillance program. Indeed, the expectation is that surveillance data will be used to develop intervention and prevention programs. Unfortunately, many PPSPs devote few resources toward such efforts.

When PPSPs implement intervention and prevention programs, they often do so without incorporating a careful strategy for evaluating efficacy. While this approach is expedient and less resource-intensive, it also makes it difficult to measure the impacts of program activities. Evaluation issues should be discussed and examined before embarking on any intervention activity. It is often advantageous, for example, to conduct a pilot intervention program that contains a component to demonstrate impact before implementing the intervention on a broader scale. To achieve this, PPSPs with limited resources may need to seek supplemental funding sources and collaboration with university or research institute partners.

9.1.1 Education

9.1.1.1 At Home and at Work

Interventions that involve education of the affected persons, their HCPs, and sometimes their employer are a common part of case investigation protocols. An example of one approach comes from an Oregon Health Division PPSP investigation of a food storage warehouse where a group of workers became ill following a pyrethrin fogging. The PPSP's site investigation identified 12 symptomatic workers, 6 of whom had sought medical care. The investigation also revealed that the warehouse had been posted indicating when reentry would be safe, but the posting failed to communicate that the warehouse should be ventilated for at least 1 hour before entry. That information had been verbally communicated by the applicator to at least one worker at the warehouse. As a result of its investigation, the PPSP provided several recommendations. It advised the employer, for instance, to develop routine procedures for mechanical ventilation after fogging treatments, and to require that the pesticide applicator provide written instructions for the timing of building opening and ventilation. Additional issues about managing possible exposures were also discussed. These recommendations were provided to the employer in writing, with a request for a response by a specific date. In his response, the employer indicated that the procedures had been implemented. This was confirmed by follow-up contact with employees. Ideally, additional follow-up should be conducted approximately 1 year later to confirm that the changes were still in place. This type of intervention and follow-up should be a routine part of all PPSP case investigation activities.

9.1.1.2 Educating the Educators

In 1993, the Oregon Health Division PPSP noted an annual pattern of multiple incidents of pesticide-related illness associated with pesticide applications in school buildings and/or school grounds. Three such incidents occurred in 1993, including one that resulted in at least 13 symptomatic persons and temporary school closure. The PARC multiagency board sent a mailing to all school district superintendents in the State containing suggested guidelines on pesticide use in schools, a checklist, and additional printed information [PARC 1993]. This letter was sent annually from 1993 to 1995, and then every other year. The mailing now incorporates additional information from the EPA on IPM in schools. This intervention has not been formally evaluated to determine whether the superintendents have adopted the guidelines or passed the information along to appropriate staff. However, illness incidents associated with applications on school property have declined to one per year since 1993 [Thomsen 2001a].

9.1.1.3 Educating the General Public

Some PPSPs conduct broad public education campaigns, especially when involved with surveillance of nonoccupational pesticide-related illness and injury. These programs may be created by PPSPs on their own, but are more commonly developed as cooperative activities with partner agencies and organizations. Campaign elements include dissemination of written materials (some formatted as fotonovelas) and use of multilingual radio programs and videos.

In addition to distributing these materials directly, PPSPs often partner with other programs to integrate pesticide safety information into diverse public health or educational efforts. Examples include child safety programs through migrant education and outreach programs for women on environmental health. Mexican consulates with Cultural Center programs have initiated traveling outreach campaigns to Mexican nationals, providing them with information about legal and social programs in the United States and Mexico. These efforts have integrated information about pesticide-related illness prevention and reporting and other occupational health issues.

Some States have distributed educational information about proper pesticide mixing and application through retail outlets that serve residential users. The Master Environmentalist program, promoted through the American Lung Association in the Northwest, addresses pesticide exposure prevention in information about household environmental hazards [Leung et al. 1997; American Lung Association of Washington 2001; EPA 2001]. Whenever possible, materials developed for individual educational presentations should be preserved in a form that allows them to be used with other audiences (e.g., PowerPoint presentations).

9.1.2 Engineering Controls, Modifications

Many types of engineering controls can be used as interventions to prevent or mitigate exposures, some of which are briefly described here. These controls can be voluntary or part of a regulatory intervention. All of these approaches are methods of reducing exposure through some form of mechanical change in the mixing, loading, or application process. Like all mechanical systems, they are effective only if properly maintained and the operators are trained in their proper use and maintenance.

9.1.2.1 Closed Mixing and Loading Systems

These are effective in reducing contact between the pesticide handler and the concentrated pesticide product. Closed systems for liquid formulations feature a probe inserted into the pesticide container which prevents fluid from spilling when transferring material to a spray tank or when connecting the pesticide container to the spray application system. Another component of closed systems are dry disconnects, which are fittings that prevent pesticide leakage when pipes or hoses accidentally become disconnected.

9.1.2.2 Creative Formulations and Technologies

Wettable powder formulations that come in dissolvable packets can significantly reduce the risk of pesticide exposure during the mixing and loading process. Newer technologies include the development of gel packs for liquid formulations. (Gel packs are water soluble packets of a liquid pesticide that has been converted to a gel material.) Other packaging and package opening systems combined with application equipment are available for some granular products.

9.1.2.3 Enclosed Cabs and Other Protection

The use of enclosed cabs for application or for workers responsible for flagging during aerial applications can serve as protection against dermal exposure. Cabs that are equipped with a ventilation system that includes a filtration device meeting ANSI/ASAE Standard 525-2 [ANSI/ASAE 1998] can be used to protect against inhalation exposure. This standard is referenced in the California Code of Regulations definition of an "Enclosed cab acceptable for respiratory protection" [California Code of Regulations 2002]. Specific performance criteria for ventilation systems in enclosed cabs appear in the EPA WPS [40 CFR Part 170.240(d)(5)]. Use of splash guards on mixing tanks may also reduce the potential for exposures.

9.1.2.4 Application and Equipment Modifications

Switching from hand application methods such as backpack or hand-held sprayer to powered spray equipment may reduce exposures. So may other modifications to material handling equipment, workstations, and ventilation in nurseries, packing sheds, greenhouses, and enclosed operations. Pesticide manufacturing and reformulation activities are also well suited to engineering controls to reduce worker exposure.

9.1.3 Administrative Controls and Regulatory Changes

The most desirable control is to substitute an equally effective but nontoxic control mechanism. Regrettably these are often not available. Collaboration with partners who are specialists in IPM may identify approaches that allow use of a less toxic compound or less frequent use of a more toxic compound. Additionally, workers can be rotated to reduce exposure to toxic chemicals that must be used.

Surveillance data may indicate that reentry intervals are inadequate for a particular pesticide in relation to a certain crop or other local conditions. Further evaluation may then lead to changes in reentry intervals. Periodic biological monitoring of workers is another method to help ensure that they are not overexposed to pesticides. The effectiveness of this intervention in reducing exposures and illness has not been well studied, however [Keifer 2000; Fillmore and Lessenger 1993]. It is not required in most States, the main exceptions being California and Washington State where cholinesterase monitoring is required for some agricultural workers.

9.1.4 Examples of Interventions

Following are examples of interventions that were developed by PPSPs to resolve emerging pesticide problems.

9.1.4.1 Automatic Insecticide Dispensers

In May 1999, the Florida Pesticide Illness Surveillance Program (PISP) identified an exposure event resulting from a restaurant's improperly placed automatic insecticide dispenser. Three persons developed illness associated with exposure to pyrethrin and piperonyl butoxide. Restaurants and other businesses use the dispensers to control indoor flying insects, usually placing them near entrances where they periodically spray pyrethrins or pyrethroids. After the PISP investigated the exposure incident, they contacted CDC/NIOSH, where surveillance data were reviewed to determine whether additional cases of illnesses were associated with automatic insecticide dispensers. Data were supplied by the Toxic Exposure Surveillance System of AAPCC, the Montana DA, the National Pesticide Telecommunication Network (now NPIC), the CDPR-PISP, and the Washington Department of Health PPSP.**

The review identified 97 cases of pesticide-related illness associated with these devices from 1986 through 1999. Three cases involved exposure to resmethrin; the rest involved the combination of pyrethrin and piperonyl butoxide. Fifty-five (57%) cases were work-related. Exposures were associated with dispensers placed too close to patron and workstation areas, dispensers placed in areas where pesticides were entrained in room air flow, and dispensers serviced by persons unfamiliar with them [CDC 2000].

This marked the first time pesticide-related illnesses attributable to automatic insecticide dispensing devices had been documented, and it brought the issue to the attention of public health officials, consumer groups, and the EPA. The report recommended nonchemical alternatives to control flying insects plus recommendations for proper installation and warning stickers whenever automatic insecticide dispensers are used. EPA has requested that the registrants of these products respond to CDC recommendations for use modification and warning labels.

9.1.4.2 Mevinphos

Mevinphos is a highly toxic organophosphate insecticide that is readily absorbed through the skin [Formoli et al. 1994]. Its high volatility at normal farm field temperatures also makes it a respiratory hazard. Events that occurred in California, Washington, and Florida contributed to the voluntary cancellation of this pesticide in 1995. In 1978, California instituted requirements for closed mixing and loading systems for mevinphos. Despite these requirements, between 1982 and 1989, there were 112 cases associated with mevinphos alone, plus another 466 cases associated with tank mixes that included mevinphos. Sixty-eight (12.6%) of the 578 cases were hospitalized. Health officials determined that a potential of exposure from small spills and inhalation still existed even with the use of closed mixing and loading systems [Formoli et al. 1994]. There were additional concerns that using PPE in hot environments could lead to heat-related illness, and even short-term removal of a respirator between loading

---

**Data from the first three sources were supplied through the EPA.

operations could result in exposures when working with this highly volatile compound.

In 1993, mevinphos was used in Washington State to address an infestation of aphids in apple and pear orchards. Phosphamidon, the chemical once commonly used to manage this pest, had been discontinued by the manufacturers. The Washington State DA (WSDA) issued emergency rules on the use of mevinphos with restrictions based on input from the manufacturer. These emergency rules included requirements for a 96-hour restricted entry period (increased from 48 hours), limits on the temperature at time of application, prohibition of hand application, and requirements for an observer during mixing and loading operations. As an additional requirement, the registrant was to make available training on safe use of the pesticide.

Despite these precautions, 26 poisonings were reported to the surveillance program within a 3-month period of use. Twenty-three of the cases (88%) were involved with mixing and loading of mevinphos, and seven of the 26 (27%) workers were hospitalized. Washington State Department of Health (WSDOH) PPSP's analysis of cases revealed that 22% of cases occurred despite compliance with all of the safety requirements. On August 19, 1993, WSDA required that only licensed applicators could mix, load, or apply mevinphos, and on August 30, 1993, they issued an emergency order to suspend the use of Phosdrin® (mevinphos) on fruit trees. It is interesting to note that although orchards in Oregon were also suffering from infestations of the same pest, no reports of mevinphos poisoning were reported during that period. Review of information from agricultural representatives indicated that Oregon growers in the affected areas had decided not to use mevinphos due to concerns about toxicity and the potential for drift from ground applications in the orchards.

The EPA had begun a data call-in to evaluate risk reduction measures needed to protect workers from mevinphos and four other organophosphate insecticides. In June 1994, the EPA considered suspending all mevinphos registrations but received a voluntary request from the only registrant to cancel all US registrations. The cancellation was granted and effective July 1, 1994, and later amended to extend the date to November 30, 1995 [EPA 2000b].

In this example, surveillance data were used to develop a regulation-mandated engineering control combined with administrative controls requiring protective equipment and procedures. Despite these changes, serious poisonings continued to occur. Prompt regulatory action prevented many additional cases. Data from California, additional information from Washington, and a report of field worker poisonings in Florida [Baer and Penzell 1993] all contributed to EPA's understanding of the hazards posed by this chemical. This information was taken into account as the chemical went through the reregistration process resulting in its voluntary cancellation in the United States. Mevinphos continues to be used in other countries.

9.1.4.3 Bis(tributyltin)oxide (TBTO) Paint Additives

This final example describes the circumstances that led to cancellation of some uses of TBTO because of concerns about nonoccupational exposures. TBTO is a fungicide commonly used as a paint additive to prevent mildew growth. Health problems from the indoor use of paints containing this pesticide have occurred in Oregon [Thomsen 1997], Washington, and Wisconsin [CDC 1991]. In all three States, investigations consisted of measurement and detection of this fungicide in the indoor environment. These findings raised concerns about acute and chronic effects of exposure and resulted in extensive efforts to remove wall materials or seal wall surfaces where the paint product was used.

WSDOH investigated an unpublished case series of six exposure incidents involving nine symptomatic persons from 1987 through 1991 [State of Washington 1987a,b,c, 1988a,b,c,d,e, 1991]. In 1988, the WSDOH issued a health warning about the use of the paint additive on interior surfaces. The manufacturer of the product involved in all of the Washington cases initiated a label change in 1987 specifying that the product should be used only for exterior applications. In July 1988, the WSDA enacted rules that prohibited both the use of paint additives, paints, and stains that contain tributyltin in inhabited structures, and the registration of products that do not clearly warn on the label that they are not for use on interior surfaces. With the exception of one case in 1991, no additional cases have occurred in Washington State. In 1992, based on the Washington investigations and a series of investigations in Oregon, the Oregon DA agreed to cancel or deny registration of TBTO paint additive products with labels that listed interior use, or did not state *for exterior use only*, and to issue stop-sale orders for those found to be in the channels of commerce. There have been no reported cases in Oregon involving use of TBTO paint additives in interior paints since 1993 [PARC 1992; Thomsen 2001b].

9.2 Evaluation of the Surveillance System

All systems should be reviewed periodically to determine whether they are effectively meeting objectives and carrying out activities efficiently. Complete guidelines for evaluating surveillance systems appear in several excellent references [CDC 2001; Romaguera et al. 2000]. Since these guidelines are readily available, this manual will not describe them in detail but will only touch on a few pertinent areas. Going through a formal evaluation of the program can highlight areas of strengths and weaknesses in the surveillance program. This information is useful for developing strategies for improvement. PPSPs are urged to schedule periodic evaluations of their programs.

The evaluation process involves a structured approach for clearly stating the goals of surveillance and determining if they are being met. If one goal of the system is to provide an estimate of the magnitude of pesticide-related illness and injury, the evaluation should determine if this goal is being met. The ability of the system to identify new emerging problems and populations at risk and to define areas for further research and studies are all PPSP functions that can be examined.

Evaluating the operational aspects of the system is a helpful tool to ensure that the system is running effectively. Operational evaluation should examine the roles of surveillance program staff, information flow, protocols for data collection and management, dissemination of findings, feedback to reporting providers and cases, and the effectiveness of interventions.

Other attributes of the surveillance system that should be evaluated include simplicity, flexibility, data quality, acceptability, sensitivity, predictive value positive, representativeness, timeliness, and stability. Because some attributes can conflict with other attributes (that is, excelling in one attribute may hamper the ability to satisfy another attribute), it is important to identify and strengthen those attributes that are most important to a particular surveillance system. It should be recognized that it may not be possible to fully achieve the less important attributes. Recommended approaches on how to assess these attributes are available [CDC 2001].

The productivity of the surveillance system should also be assessed. This includes determining (1) the number of investigative reports that are issued, (2) the number of case reports received and followed up, (3) the number of interviews conducted, (4) the number of site inspections performed, (5) the number of phone consultations performed, and (6) the frequency of summary surveillance reports. A summary of outreach activities to stakeholders should also be prepared (e.g., number of newsletter articles published and the number of presentations delivered). Having an electronic tracking system can facilitate the collection and evaluation of this information.

Conclusions and recommendations should be provided when evaluating a surveillance system. An assessment should be made about whether the surveillance system should be continued (that is, can justification be made for the resources applied to it?). If the answer is yes, the need for any modifications to the system should be identified. Finally, when making recommendations for modifications, it is prudent to recall that the costs and attributes of the system are interdependent (e.g., improvements in one attribute may increase costs or affect the performance of another attribute). Therefore, these consequences should be considered when recommending modifications. An evaluation should examine whether recommendations made by the PPSP to prevent pesticide poisoning were implemented, or if other actions were taken as a result of surveillance data (e.g., are data used for generating research or interventions?). If the PPSP has an advisory committee (see Section 5.9.3), this committee can often contribute to the evaluation process.

9.3 Conclusion

The strategies and examples described in this chapter demonstrate how surveillance data can be used to identify new emerging pesticide problems, and thus estimate the magnitude of pesticide poisoning.

Throughout this instruction guide, our goal has been to provide tools that NIOSH partners can use to identify pesticide poisoning risk factors and target interventions toward them. Those tools include

- Information about the importance of pesticide poisoning surveillance
- Mechanisms to improve reporting of cases to surveillance programs
- Methods to investigate reported cases
- Guidance on using the case definition, and
- Additional resources relevant to pesticide poisoning surveillance.

They are based on the experience of public health professionals who are directly involved in the field. Developed over many years, these tools can provide a strong base for any State or local entity initiating a pesticide surveillance program.

It is our hope that others will continue to build on this body of knowledge and expand on the surveillance activities being performed today. In particular, we look forward to a repository of successful intervention and prevention strategies and an examination of their ability to reduce the occurrence of pesticide related-illness and injury. Much remains to be done in assessing the effectiveness of IPM techniques, administrative controls, and the use of low-toxicity pesticides when nontoxic methods are impractical or unsuccessful.

References

References

Alexander RW, Fedoruk MJ [1986]. Epidemic psychogenic illness in a telephone operators' building. J Occup Med 28:42–45.

American Lung Association of Washington [2001]. MHE program: is your home healthy for you and your children? [http://www.alaw.org/air_quality/information_and_referral/master_home_environmentalist/] Date accessed: May 10, 2001.

AMA (American Medical Association) [1997]. Educational and informational strategies to reduce pesticide risks. Prev Med 26:191–200.

Ames RG, Brown SK, Mengle DC, Kahn E, Stratton JW, Jackson RJ [1989]. Cholinesterase activity depression among California pesticide applicators. Am J Ind Med 15:143–150.

ANSI/ASAE [1998]. Agricultural cabs–environmental air quality. Part 2: Pesticide vapor filters–test procedure and performance criteria. New York: American National Standards Institute, ANSI/ASAE Standard S525-2

Baer RD, Penzell D [1993]. Research report: susto and pesticide poisoning among Florida farmworkers. Cult Med Psychiatry 17:321–327.

Baum L [2001a] Personal communication of March 12, 2001, between L. Baum, Washington Department of Health, and M. Barnett, Strategic Options Consulting, Inc.

Baum L [2001b] Personal communication of March 29, 2001, between L. Baum, Washington Department of Health, and M. Barnett, Strategic Options Consulting, Inc. Capture-recapture exploratory analysis.

Bernard HR [1995]. Research methods in anthropology: qualitative and quantitative approaches. 3rd ed. Walnut Creek, CA: Alta Mira Press.

California Code of Regulations (CCR) [2002]. Title 3. California Code of Regulations; Division 6. Pesticides and Pest Control Operations; Chapter 1. Pesticide regulatory program; Subchapter 1. Definitions of terms; Article 1. Definitions for Division 6.

California EPA [2002]. Guidelines for physicians who supervise workers exposed to cholinesterase-inhibiting pesticides. 4th ed. Oakland, CA: California Environmental Protection Agency, Office of Environmental Health Hazard Assessment.

Calvert GM, Sanderson WT, Barnett M., Blondell JM, Mehler LN [2001]. Surveillance of pesticide-related illness and injury in humans. In: Krieger R, ed. Handbook of pesticide toxicology. 2nd ed. San Diego, CA: Academic Press.

Calvert GM, Mehler LN, Rosales R, Baum L, Thomsen C, Male D, Shafey O, Das R, Lackovic M, Arvizu E [2003]. Acute pesticide-related illness among working youth, 1988B1999. Am J Public Health 93:605–610.

Calvert GM, Plate DK, Das R, Rosales R, Shafey O, Thomsen C, Male D, Beckman J, Arvizu E, Lackovic M [2004]. Acute occupational pesticide-related illness in the US, 1998–1999: surveillance findings from the SENSOR-Pesticides Program. Am J Ind Med 45:14–23.

CDC (Centers for Disease Control and Prevention) [1991]. Acute effect of indoor exposure to paint confining bis(tributyltin) oxide. MMWR 40(17):280–281.

CDC [2000a] Surveillance in a suitcase. Atlanta, GA: Centers for Disease Control and Prevention, Epidemiology Program Office. [http://www.cdc.gov/epo/surveillancein] Date accessed: October 2004.

CDC (Centers for Disease Control and Prevention) [2000b]. Illnesses associated with use of automatic insecticide dispenser units–selected States and United States, 1986–1999. MMWR 49(22):492–495.

CDC (Centers for Disease Control and Prevention) [2001]. Updated guidelines for evaluating public health surveillance systems. MMWR 50(RR–13):1–35.

CDC (Centers for Disease Control and Prevention) [2003]. HIPAA privacy rule and public health: guidance from CDC and the U.S. Department of Health and Human Services. MMWR 52(Suppl):1–20.

CDPR [2002]. Pesticide Illness Surveillance Program, 1993–96. Sacramento, CA: California Department of Pesticide Regulation. Database.

CFR. Code of Federal regulations. Washington, DC: U.S. Government Printing Office, Office of the Federal Register.

Cole TB, Chorba TL, Horan JM [1990]. Patterns of transmission of epidemic hysteria in a school. Epidemiol 1(3):212–218.

CSTE [1996]. CSTE position statement 1996–15: adding acute pesticide poisoning/injuries (APP/I) as a condition reportable to the National Public Health Surveillance System (NPHSS). Atlanta, GA: Council of State and Territorial Epidemiologists.

CSTE [1999]. CSTE position statement ENV–3: inclusion of acute pesticide-related illness and injury indicators in the National Public Health Surveillance System (NPHSS). Atlanta, GA: Council of State and Territorial Epidemiologists.

Department of Commerce [1980]. Standard occupational classification manual, 1980. Washington, DC: U.S. Department of Commerce, Office of Federal Statistical Policy and Standards.

Donaldson D, Kiely T, Grube A [2002]. Pesticides industry sales and usage. 1998 and 1999 market estimates. Washington DC: U.S. Environmental Protection Agency, Report No. EPA–733–R–02–001.

EPA [1996]. EPA completes cleanup of toxic pesticide in Detroit homes. Washington, DC: U.S. Environmental Protection Agency, EPA Press Release No. 96–OPA090, May 7, 1996. [http://www.epa.gov/reg5oopa/news/96/opa90.htm].

EPA [1997]. Inert ingredients in pesticide products; policy statement. Washington, DC: U.S. Environmental Protection Agency, OPP–36140; FRL–3190–1.

EPA [2000a]. Manual of manuals. Summaries and ordering information for eight laboratory analytical chemistry methods manuals published by the former Environmental Monitoring Systems Laboratory-Cincinnati (EMSL-Cincinnati) between 1988 and 1995. Washington, DC: U.S. Environmental Protection Agency. [http://www.epa.gov/nerlcwww/methmans.html] Date accessed: October 2004.

EPA [2000b]. Mevinphos: revised human health risk assessment. Chemical I.D. No. 015801. Washington, DC: U.S. Environmental Protection Agency, Office of Prevention, Pesticides, and Toxic Substances, Health Effects Division, Reregistration Branch 1.

EPA [2001]. Administrator launches Portland's Master Home Environmentalist Program. Washington, DC: U.S. Environmental Protection Agency, EPA News Release No. 01–035. [http://yosemite.epa.gov/r10/homepage.nsf/0/4df80ee1706b1b1488256aea005fdfe4?OpenDocument] Date accessed: October 2004.

EPA [2002]. Federal Insecticide, fungicide, and rodenticide act (FIFRA) inspection manual. Washington, DC: U.S. Environmental Protection Agency, Office of Enforcement and Compliance Assurance, Office of Compliance, EPA 305—02–001.

EPA/NIOSH [1991]. Building air quality: a guide for building owners and facility managers. EPA Publication No. 400/1–91/003, DHHS (NIOSH) Publication No. 91–114.

52 Fed. Reg.13305 [1987]. EPA: inert ingredients in pesticide products; policy statement.

64 Fed. Reg. 112 [1999]. EPA: inert ingredients no longer used in pesticide products.

Fillmore CM, Lessenger JE [1993]. A cholinesterase testing program for pesticide applicators. J Occup Med; 35:(1)61–70.

Formoli T, Thongsinthusk T, Sanborn, J [1994]. Estimation of exposure of persons in California to pesticide products that contain mevinphos. Sacramento, CA: California Environmental Protection Agency, Department of Pesticide Regulation Worker Health and Safety Branch, Publication HS–1653.

GAO [1994]. Pesticides on farms. Limited capability exists to monitor occupational illnesses and injuries. Washington, DC: U.S. General Accounting Office, GAO/PEMD–94–6.

GAO [1999]. Pesticides. Use, effects, and alternatives to pesticides in schools. Washington, DC: U.S. General Accounting Office, GAO/RCED–00–17.

GAO [2000]. Pesticides: improvements needed to ensure the safety of farmworkers and their children. Washington, DC: U.S. General Accounting Office, GAO/RCED–00–40

Green M, Heumann M, Sokolow R, Foster LR, Bryant R, Skeels M [1990]. Public health implications of the microbial pesticide Bacillus thuringiensis: an epidemiologic study, Oregon 1985–86. Am J Public Health 80(7):848–852.

Guidotti TL, Alexander RW, Fedoruk MJ [1987]. Epidemiologic features that may distinguish between building-associated illness outbreaks due to chemical exposure or psychogenic origin. J Occup Med 29(2)148–150.

Hartge P, Cahill J [1998] Field methods in epidemiology. In: Rothman KJ, Greenland S, eds. Modern epidemiology. 2nd ed. Philadelphia, PA: Lippincott-Raven.

Heumann M [2000]. Personal conversation on October 30, 2000, between M. Heumann, Oregon Health Division, and Margot Barnett, Strategic Options Consulting, Inc.

Keifer MC [2000] Effectiveness of interventions in reducing pesticide overexposure and poisonings. Am J Prev Med 18(4S):80–89.

Leung R, Koenig JQ, Simcox N, van Belle G, Fenske R, Gilbert SG [1997]. Behavioral changes following participation in a home health promotional program in King County, Washington. Environ Health Perspec 105(10):1132–1135.

Maizlish NA, ed. [2000]. Workplace health surveillance. New York: Oxford University Press.

NAS [2000]. The future role of pesticides in US agriculture. Washington, DC: National Academy Press.

NCHS [2003]. Instruction manual, part 19b: alphabetical index of industries and occupations, 2003. Hyattsville, MD: National Center for Health Statistics.

New Jersey Department of Health [2000]. Occupational Disease reporting requirements–hospitals. Trenton, NJ: Department of Health and Human Services, Division of Epidemiology and Environmental and Occupational Health. (Adopted 9/18/00.)

New York State Department of Health [1997]. SPIDER user's guide. Albany, NY: New York State Department of Health.

NIOSH [1994]. NIOSH manual of analytical methods. 4th ed. Cincinnati, OH: U.S. Department of Health and Human Services, Public Health Service, Centers for Disease Control and Prevention, National Institute for Occupational Safety and Health, DHHS (NIOSH) Publication No. 94–113.

NIOSH [2004]. Case definition for acute pesticide-related illness and injury cases reportable to the national public health surveillance system. Cincinnati, OH: U.S. Department of Health and Human Services, Public Health Service, Centers for Disease Control and Prevention, National Institute for Occupational Safety and Health. [http://www.cdc.gov/niosh/topics/pesticides/].

OMB [1987]. Standard industrial classification manual, 1987. Washington, DC: Executive Office of the President, Office of Management and Budget.

OMB [1997]. North American industry classification system, United States, 1997. Washington, DC: Executive Office of the President, Office of Management and Budget.

OMB [2000]. Standard occupational classification manual, 2000. Washington, DC: Executive Office of the President, Office of Management and Budget.

OSHA [2000]. Index of sampling and analytical methods. Washington, DC: U.S. Department of Labor, Occupational Safety and Health Administration.[http://www.osha-slc.gov/dts/sltc/methods/toc.html] Date accessed: October 2004.

PARC [1992]. Pesticide analytical and response center annual report for 1991–92. State of Oregon.

PARC [1993]. Oregon pesticide analytical and response center, 1993 annual report. Portland, OR: Oregon Health Division.

Pearce M, Behie G, Chappell N [2002]. The effects of aerial spraying with Bacillus thuringiensis kurstaki on area residents. Environ Health Rev 02.1:19–22. [http://www.ciphi.ca/ehr/EHR02.1-19-22.pdf] Date accessed: October 2004.

Reigart JR, Roberts JR, eds. [1999]. Recognition and management of pesticide poisonings. 5th ed. Washington, DC: U.S. Environmental Protection Agency, Office of Pesticide Programs, EPA 735–98–003.

RCW. (Revised Code of Washington) [http://slc.leg.wa.gov/default.htm] Date accessed: October 2004.

Romaguera RA, German RR, Klaucke DN [2000]. Evaluating public health surveillance. In: Teutsch SM, Churchill RE, eds. Principles and practice of public health surveillance. 2nd ed. New York: Oxford University Press.

Rothman KJ, Greenland S, eds. [1998]. Modern epidemiology. 2nd ed. Philadelphia, PA: Lippincott-Raven Publishers.

Spinello E [2001]. Introduction to the use of GIS in public health. [http://www.pitt.edu/~superl1/lecture/lec1291/] Date accessed: October 2004.

State of Washington [1987a]. Pesticide incident summary report: bis (tri-butyltin) oxide. State of Washington: Department of Health, DOH Case No. 870528. Unpublished.

State of Washington [1987b]. Pesticide incident summary report: bis (tri-butyltin) oxide. State of Washington: Department of Health, DOH Case No. 870617. Unpublished.

State of Washington [1987c]. Pesticide incident summary report: bis (tri-butyltin) oxide. State of Washington: Department of Health, DOH Case No. 871006. Unpublished.

State of Washington [1988a]. Pesticide incident summary report: bis (tri-N-butyltin) oxide. State of Washington: Department of Health, DOH Case No. 880102. Unpublished.

State of Washington [1988b]. Pesticide incident summary report: bis (tributyltin) oxide. State of Washington: Department of Health, DOH Case No. 880307. Unpublished.

State of Washington [1988c]. Chemical analysis: paint. State of Washington: Department of Health, Office of Toxic substances, Case No. 8722042 and 8722043. Unpublished.

State of Washington [1988d]. Health advisory issued on use of TBT in interior house paints. State of Washington: Department of Social and Health Services, Press Release No. 88–50.

State of Washington [1988e]. Pesticide incident summary report: bis (tri-butyltin) oxide. State of Washington: Department of Health, DOH Case No. 871006. Unpublished.

State of Washington [1991]. Pesticide incident summary report: bis (tri-butyltin) oxide. State of Washington: Department of Health, DOH Case No. 91–1186. Unpublished.

Teutsch SM, Churchill RE, eds. [2000]. Principles and practice of public health surveillance. 2nd ed. New York: Oxford University Press.

Thomsen C [1997]. Telephone conversation on October 24, 1997, between C. Thomsen, Oregon Health Division, and M. Barnett, Strategic Options Consulting, Inc.

Thomsen, C [2001a]. Telephone conversation on April 19, 2001, between C. Thomsen, Oregon Health Division, and M. Barnett, Strategic Options Consulting, Inc.

Thomsen C [2001b]. Telephone conversation on May 17, 2001, between C. Thomsen, Oregon Health Division, and M. Barnett, Strategic Options Consulting, Inc.

Washington State Department of Labor and Industries [1995]. Cholinesterase monitoring in Washington State: report from the Technical Advisory Group.

Washington State Department of Labor and Industries [2003]. Public meetings scheduled for pesticide rules. L&I News. [http://www.lni.wa.gov/news/2002/pr020703a.htm] Date accessed: October 2004.

Washington State Department of Labor and Industries [2003]. Safety Standards for Agriculture; Chapter 296–307 WAC (Washington Administrative Code), Part J–1, Cholinesterase monitoring.

Washington State Legislature [2002]. Washington Administrative Code. 97–09–013, recodified as '296 307A–14520 and 96–22–048, '296–306A–14520. What are the department's recommendations for cholinesterase monitoring? (Nonmandatory) [http://www.leg.wa.gov/wac/index.cfm?fuseaction=Section&Section=296-307-14520] Date accessed: October 2004.

WHO [1977]. International classification of diseases: manual of the international statistical classification of diseases, injuries, and causes of death. Vol. 1. Geneva, Switzerland: World Health Organization.

WHO [1992]. International statistical classification of diseases and related health problems (ICD–10). Geneva, Switzerland: World Health Organization.

Appendix A
Listing of Morbidity and Mortality Weekly Report (MMWR) Articles on Pesticide-Related Illness and Injury January 1, 1982–September 30, 2004

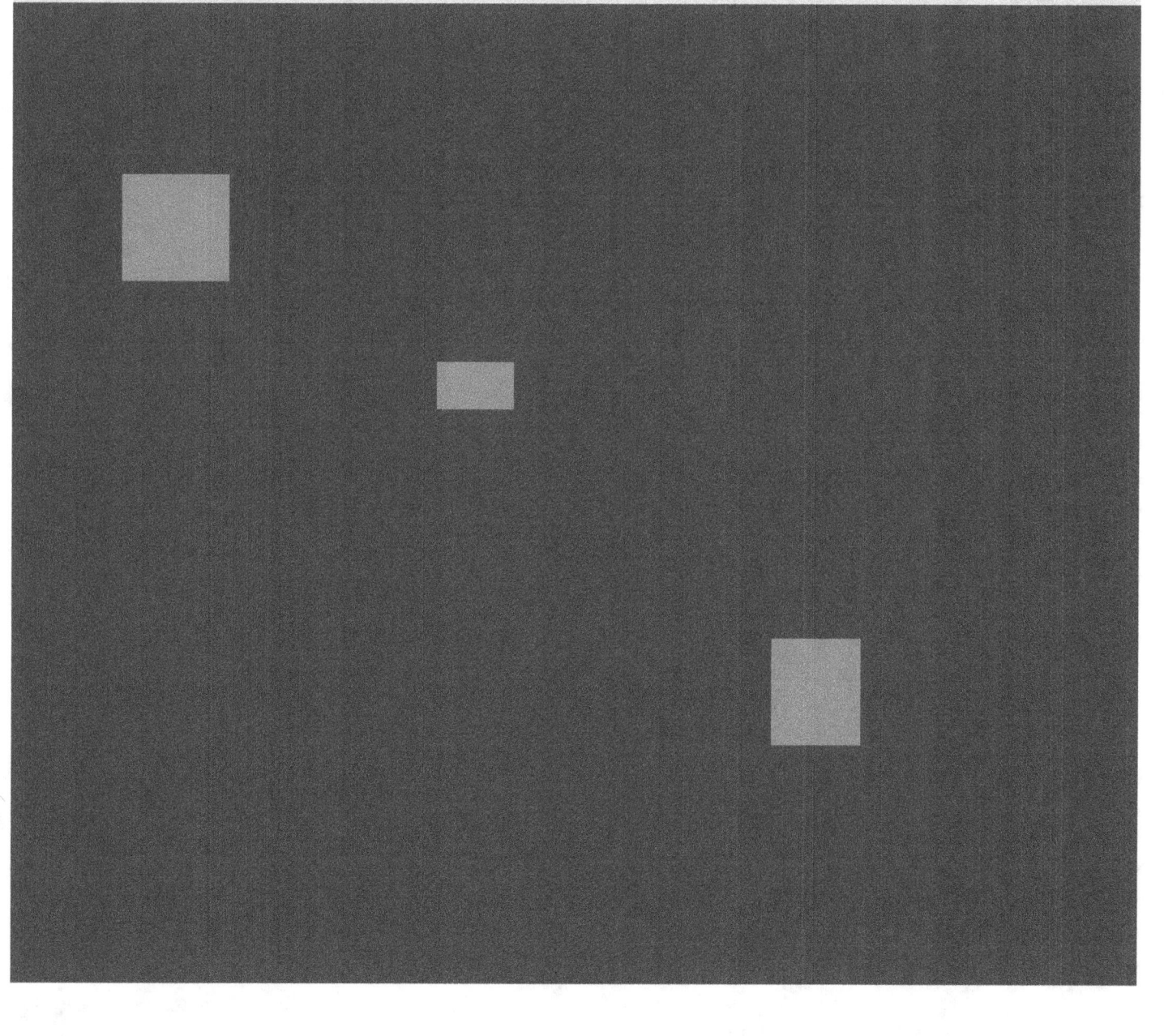

Appendix A ■

Listing of Morbidity and Mortality Weekly Report (MMWR) Articles on Pesticide-Related Illness and Injury January 1, 1982–September 30, 2004

Note: This listing may be incomplete; efforts were made to include all articles that addressed this topic, but some may have been missed by the search strategy used. The list is chronological with most recent articles listed first. Relevant Surveillance Summaries and Recommendations and Reports are listed separately.

Illness associated with drift of chloropicrin soil fumigant into a residential area – Kern County, California, 2003. Vol. 53, No. 32: 740–742, 8/20/2004.
http://www.cdc.gov/mmwr/preview/mmwrhtml/mm5332a4.htm

Surveillance for acute insecticide-related illness associated with mosquito-control efforts—Nine states, 1999–2002. Vol. 52, No. 27:629–634, 07/11/2003.
http://www.cdc.gov/mmwr/preview/mmwrhtml/mm5227a1.htm

Nicotine poisoning after ingestion of contaminated ground beef—Michigan, 2003. Vol. 52, No. 18:413–416, 05/09/2003. http://www.cdc.gov/mmwr/preview/mmwrhtml/mm5218a3.htm

Poisoning by an illegally imported Chinese rodenticide containing tetramethylenedisulfotetramine—New York City, 2002. Vol. 52, No. 10:199–201, 03/14/2003.
http://www.cdc.gov/mmwr/preview/mmwrhtml/mm5210a4.htm

Pesticide-related illnesses associated with the use of a plant growth regulator—Italy, 2001. Vol. 50, No. 39:845–847, 10/05/2001. http://www.cdc.gov/mmwr/PDF/wk/mm5039.pdf

Nosocomial poisoning associated with emergency department treatment of organophosphate toxicity—Georgia, 2000. Vol. 49, No. 51:1156–1158, 01/05/2001.
http://www.cdc.gov/mmwr/PDF/wk/mm4951.pdf

Occupational fatalities associated with 2,4-Dichlorophenol (2,4-DCP) exposure, 1980–1998. Vol. 49, No. 23:516–518, 06/16/2000. http://www.cdc.gov/mmwr/PDF/wk/mm4923.pdf

Illnesses associated with use of automatic insecticide dispenser units—selected States and United States, 1986–1999. Vol. 49, No. 22:492–495, 6/09/2000.
http://www.cdc.gov/mmwr/PDF/wk/mm4922.pdf

Surveillance for acute pesticide-related illness during the Medfly Eradication Program—Florida, 1998. Vol. 48, No. 44:1015–1018, 1027, 11/12/1999.
http://www.cdc.gov/mmwr/PDF/wk/mm4844.pdf

Illnesses associated with occupational use of flea-control products—California, Texas, and Washington, 1989–1997. Vol. 48, No. 21:443–447, 06/04/1999. http://www.cdc.gov/mmwr/PDF/wk/mm4821.pdf

Aldicarb as a cause of food poisoning—Louisiana, 1998. Vol. 48, No. 13:269–271, 04/09/1999. http://www.cdc.gov/mmwr/PDF/wk/mm4813.pdf

Outbreaks of gastrointestinal illness of unknown etiology associated with eating burritos— United States, October 1997–October 1998. Vol. 48, No. 10:210–213, 03/19/1999. http://www.cdc.gov/mmwr/PDF/wk/mm4810.pdf

Farmworker illness following exposure to carbofuran and other pesticides. Vol. 48, No. 06:113–116, 02/19/1999. http://www.cdc.gov/mmwr/PDF/wk/mm4806.pdf

Monitoring environmental disease—United States, 1997. Vol. 47, No. 25:522–525, 07/03/1998. http://www.cdc.gov/mmwr/PDF/wk/mm4725.pdf

Poisonings associated with illegal use of aldicarb as a rodenticide—New York City, 1994–1997. Vol. 46, No. 41:961–963, 10/17/1997. http://www.cdc.gov/mmwr/PDF/wk/mm4641.pdf

Acute pesticide poisoning associated with use of a sulfotepp fumigant. Vol. 45, No. 36:780–782, 09/13/1996. http://www.cdc.gov/mmwr/PDF/wk/mm4536.pdf

Pentachlorophenol poisoning in newborn infants—St. Louis, Missouri, April-August 1967. Vol. 45, No. 25:545–549, 06/28/1996. http://www.cdc.gov/mmwr/PDF/wk/mm4525.pdf

Eye injuries to agricultural workers—Minnesota, 1992–1993. Vol. 44, No. 18:364–366, 05/12/1995.

Deaths associated with exposure to fumigants in railroad cars—United States. Vol. 43, No. 27:489–491, 07/15/1994. http://www.cdc.gov/mmwr/PDF/wk/mm4327.pdf.

Occupational pesticide poisoning in apple orchards—Washington, 1993. Vol. 42, No. 51:993–995, 01/07/1994. http://www.cdc.gov/mmwr/PDF/wk/mm4251.pdf

Acute effect of indoor exposure to paint containing bis(tributyltin)oxide—Wisconsin, 1991. Vol. 40, No. 17:280–281, 05/ 03/1991.

Unintentional methyl bromide gas release—Florida, 1988. Vol. 38, No. 51:880-882, 01/05/1990.

Mercury exposure from interior latex paint—Michigan. Vol. 39, No. 8:125–126, 03/02/1990.

Endrin poisoning associated with taquito ingestion—California. Vol. 38, No. 19:345–347, 05/19/1989.

Organophosphate toxicity associated with flea-dip products—California. Vol. 37, No. 21:329–330, 335–336, 06/03/1988.

Serum 2,3,7,8-Tetrachlorodibenzo-p-dioxin levels in air force health study participants—Preliminary Report. Vol. 37, No. 20:309–311, 05/27/1988.

Fatalities resulting from sulfuryl fluoride exposure after home fumigation—Virginia. Vol. 36, No. 36:602–604, 609–611, 09/18/1987.

Serum dioxin in Vietnam-Era Veterans—Preliminary Report. Vol. 36, No. 28:470–475, 07/24/1987.

Current trends postservice mortality among Vietnam Veterans. Vol. 36, No. 05:61–64, 02/13/1987.

Aldicarb food poisoning from contaminated watermelons—California. Vol. 35, No. 16:254–258, 04/25/1986.

Acute poisoning following exposure to an agricultural insecticide—California. Vol. 34, No. 30:464–466, 471, 08/02/1985.

Neurologic findings among workers exposed to fenthion in a veterinary hospital—Georgia. Vol. 34, No. 26:402–403, 07/05/1985.

Vietnam Veterans' risks for fathering babies with birth defects. Vol. 33, No. 32:457–459, 08/17/1984.

Organophosphate insecticide poisoning among siblings—Mississippi. Vol. 33, No. 42:592–594, 10/26/1984.

DDT exposures in a natural history museum—Colorado. Vol. 32, No. 34:443–444, 09/02/1983.

Arsenic contamination in an abandoned building—Ohio. Vol. 31, No. 39:531–532, 10/08/1982.

Surveillance Summaries

Surveillance for emergency events involving hazardous substances—U.S., 1990–1992. Vol. 43, No. SS-2, 07/22/1994. http://www.cdc.gov/mmwr/PDF/SS/SS4302.pdf

Surveillance for occupational asthma—Michigan and New Jersey, 1988–1992. Vol. 43, No. SS-1:9-16, 06/10/1994. http://www.cdc.gov/mmwr/PDF/SS/SS4301.pdf

Recommendations and Reports

Diagnosis and management of foodborne illnesses a primer for physicians. Vol. 50, No. RR-2, 01/26/2001.

Biological and chemical terrorism: strategic plan for preparedness and response. Vol. 49, No. RR-4, 04/21/2000. http://www.cdc.gov/mmwr/PDF/RR/RR4904.pdf

Recommended framework for presenting injury mortality data. Vol. 46, No. RR-14:1, 08/29/1997. http://www.cdc.gov/mmwr/PDF/RR/RR4614.pdf

Mandatory reporting of occupational diseases by clinicians. Vol. 39, No. RR-9:019, 06/22/1990.

Appendix B
Selected Pesticide Illness Reporting Statutes and Rules

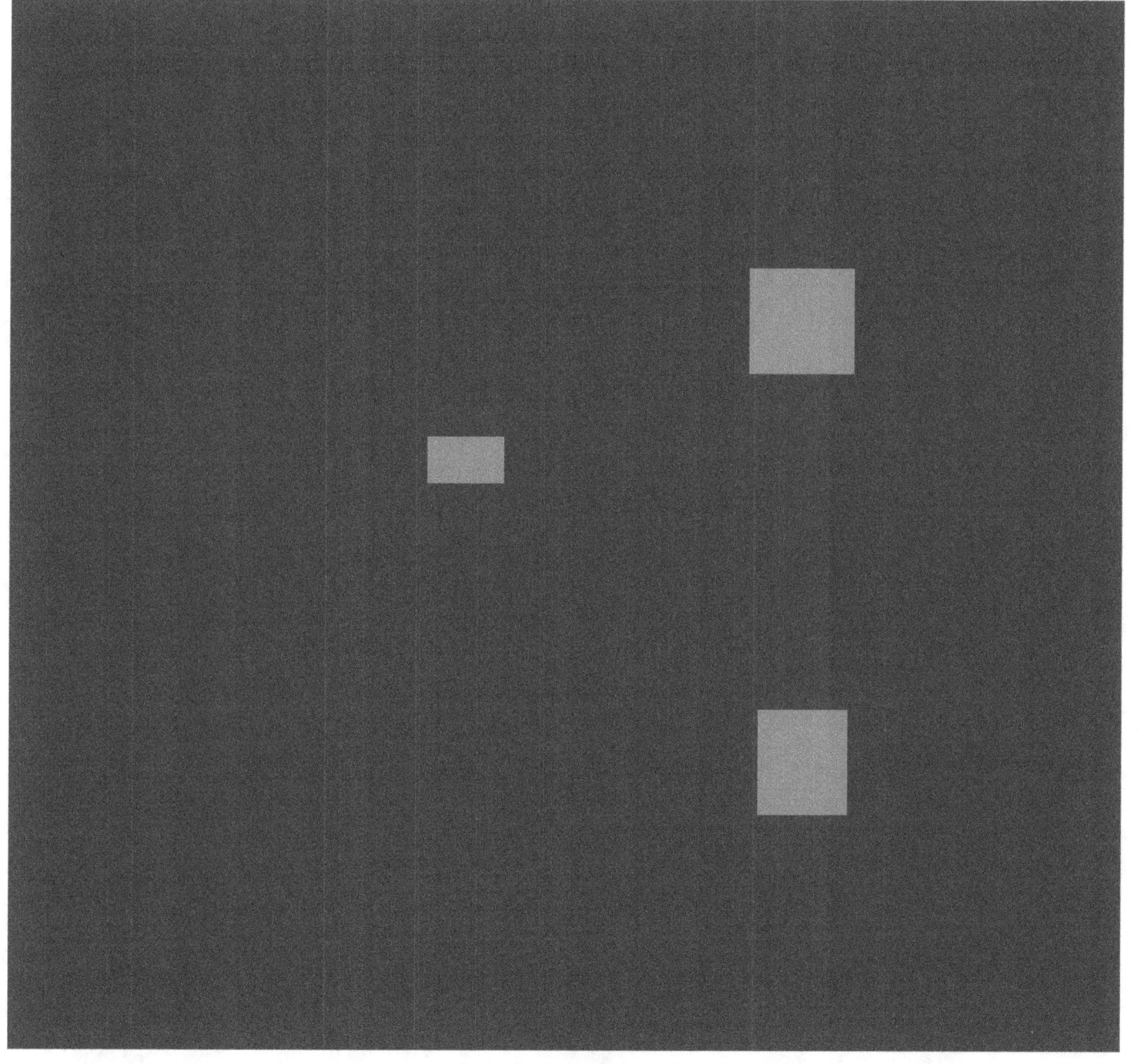

Appendix B ∎

Selected Pesticide Illness Reporting Statutes and Rules

Note: This list was current as of September 30, 2004. The list is not a comprehensive list of all applicable State reporting rule links but is intended to supply links to a rule or statute that may be of particular interest.

Arizona

http://www.azhs.gov. This page has a link to the reporting statute and rule.

California

http://www.leginfo.ca.gov/cgi-bin/calawquery?codesection=hsc.
Sections 105200–105225 of the health and safety code relate to pesticide illness reporting.

Michigan

http://www.legislature.mi.gov/mileg.asp?page=getObject&objName=mcl-368-1978-5-56&highlight.

Missouri

http://www.sos.mo.gov/adrules/csr/current/19csr/19c20-20.pdf.

New Mexico

http://www.nmcpr.state.nm.us/nmac/_title07/T07C004.htm.
Browse to *Control of disease and conditions of public health significance*.

New Jersey

http://www.state.nj.us/health/eoh/survweb/.
Browse to *occupational disease reporting requirements*.

New York

http://www.health.state.ny.us/nysdoh/environ/pest/appenda.htm.

OREGON

http://www.leg.state.or.us/ors/433.html.
Browse to *Chapter 433.004*.

TEXAS

http://lamb.sos.state.tx.us/tac/index.html.
This is a link to the Texas administrative code. Enter the TAC Viewer and select *Title 25—Health Services, Part 1, Chapter 99—Occupational Diseases.*

http://www.capitol.state.tx.us/statutes/statutes.html.
This link connects you with the Texas statutes. Select *Health and Safety Code*; search for *Chapter 84—Reporting of Occupational Conditions.*

WASHINGTON

http://search.leg.wa.gov/pub/textsearch/default.asp.
The Washington law establishing the Pesticide Incident Response and Tracking (PIRT) Board is at this site. Seach for *RCW 70.104*, after checking the *RCW & Depositions* box.

The current reporting rules can be found by searching for *WAC 246* after checking the *WAC* box. Pesticide poisoning is listed as one of the notifiable conditions in Chapter 246-101.

Appendix C
Sample Forms, Investigation Tools, and Templates for Data Tables

APPENDIX C ■

SAMPLE FORMS, INVESTIGATION TOOLS, AND TEMPLATES FOR DATA TABLES

CONTENTS

C.1   Case Tracking Form and Contact Log

C.2   Main Pesticide Exposure Questionnaire

C.3   Pesticide Illness and Injury Surveillance Data Collection Form

C.4   Field Investigation Contact Form and Health Safety Checklist for Field Personnel

C.5   Instructions for National Transportation Safety Board (NTSB) Search to Obtain Reports of Airplane Accidents Involving Aerial Pesticide Applicators

C.6   Sample Templates for Tables Presenting Surveillance Data

C.7   Sample Letters for PPSP Case Follow-up

C.8   Instructions for Obtaining Acute Pesticide-Related Illnesses and Injuries data from Poison Control Centers (PCC)

C.1 CASE TRACKING FORM AND CONTACT LOG

These two forms are examples of mechanisms that PPSPs can use to track the case follow-up process to ensure that all needed information is collected and that appropriate referrals and agency contacts are made. As noted in Chapter 5, the program should determine if it wants to log and track informational calls and/or reports that are screened out as unrelated to pesticide exposure. The tracking checklist form included here does not include informational calls but could be adapted to do so. The procedures used for any tracking system should be documented in a procedure manual. The contact log is a tool for recording names of individuals and the dates they were contacted as part of the follow-up investigation process. The comments should be supplemented with additional records of conversation pertinent to the investigation. The forms can be useful tools to review timeliness and completeness of the investigation process.

APPENDIX C ■ SECTION 1

PESTICIDE CASE TRACKING CHECKLIST

CASE ID NO. _____   EVENT ID NO. _____

		DATE	INITIALS	COMMENTS (specify names)
1. Reported to HD and Logged				
2. Data Collection Form Started				
3. Other Agency/ies Notified *(Indicate Permission to refer? Y / N)*		DATE	PERMISSION	
	Ag			
	OSHA			
	Forestry			
	EPA			
		DATE		
4. Medical Records Requested				
5. Medical Records Received				
6. Case Report Written/Filed				
7. Memo Distributed (stamped medical/confidential)				
8. Other Agency Reports Received		DATE	PERMISSION	
	Ag			
	OSHA			
	Forestry			
	EPA			
9. Case Classification		DATE		
10. Case Data Entered into SPIDER				

CASE CONTACT LOG				
Date	Contact	Phone No.	Initial	Comments

APPENDIX C ■ SECTION 2
PESTICIDE EXPOSURE QUESTIONNAIRE

C.2 MAIN PESTICIDE EXPOSURE QUESTIONNAIRE

INSTRUCTIONS

This is a sample questionnaire for use by an acute pesticide-related illness and injury surveillance program. It includes questions that satisfy the data requirements for all of the core variables needed by the National Institute for Occupational Safety and Health (NIOSH), Centers for Disease Control and Prevention (CDC). Additional questions for administrative report management at the State level, as well as optional suggested questions, are included. Optional questions are indicated on the form by framing with a dashed border. The order of the questions is designed to provide ease of data collection as well as data entry using the SENSOR Pesticide Incident Data Entry and Reporting (SPIDER) data management software. Shading on pages 1-17 indicates data to be completed by the interviewer and not asked during the actual interview. Pages 18-21 are to be completed following the initial interview as additional medical information is collected and case closure is completed. States will need to customize this questionnaire for their specific needs. Some States may choose to develop separate questionnaires for agricultural, occupational, nonoccupational, physician, or non-English speaker interviews. The design presented here is not appropriate for interviewing non-English speaking farmworkers. An example of a Spanish language questionnaire specific to agriculture situations can be obtained from the California DHS SENSOR Pesticide Poisoning California (SPPC) Program (510-620-5757 or http://www.dhs.ca.gov/ohb/AgInjury/).

APPENDIX C ■ SECTION 2
FORM CONTAINS CONFIDENTIAL INFORMATION

Pesticide Illness and Injury Surveillance Questionnaire

Case ID __ __ __ __ __ __ __ __
Event ID __ __ __ __ __ __ __ __

Interviewer Name _____

Interviewer ID __ __ __   Today's date __ __ / __ __ / __ __
(mm/dd/yy)

MAIN PESTICIDE EXPOSURE QUESTIONNAIRE

Hello. May I speak to Mr./Ms. _____ This is _____ with the _____ Department of Health. We recently received notification by _____ who you may have experienced a pesticide exposure. We try to keep track of persons in our State that have been exposed to pesticides, and what has happened to them. If you have a few moments, I would like to ask you some questions. The information you provide may help us prevent similar pesticide exposures in the future. Your participation is voluntary, and you may skip any questions you do not want to answer. Shall we begin? *(If the subject agrees, begin the interview. If the subject says he/she does not have time right now, try to schedule a time when you may call back. If the subject is unwilling to answer any questions, thank him/her for his/her time, hang up, and complete as much information as possible based on the original report.)*

Case Information Screen

First, I would just like to ask you a few questions about yourself.
(It is not necessary to ask these questions if you already have this information, although it is helpful to make sure the information is correct by reading the spelling of names and checking the address and phone number.)

1. What is your last name? _____
2. What is your first name? _____
3. What is your middle name? _____

3a. What is your Social Security Number? __ __ __ – __ __ – __ __ __ __

4. What is your home address? _____
   City _____ State __ __   Zip __ __ __ __ __
5. What is your home telephone number? (__ __ __) __ __ __ – __ __ __ __
6. What county do you live in? _____
7. Were you living in a different residence at time of exposure? ☐ Yes ☑ No

   If the person answers yes, go to Number 8; if the person answers no, skip Number 8.

8. What was your home address at the time you were exposed? *(Enter in exposure incident screen if this was location where exposure occurred.)*

   Address line 1 _____
   Address line 2 _____
   City _____
   State __ __   ZIP __ __ __ __ __   County Name _____
   FIPS code __ __ __

Rev. 7/1/04

APPENDIX C ■ SECTION 2
PESTICIDE EXPOSURE QUESTIONNAIRE

Pesticide Illness and Injury Surveillance Questionnaire

Case ID __ __ __ __ __ __ __ __
Event ID __ __ __ __ __ __ __ __

9. Gender ☐ 1 Male ☐ 2 Female ☐ 8 Other ☑ 9 Unknown

10. What is your race? ☐ 1 Am Indian ☐ 2 Asian/Pacific Is. ☐ 3 Black ☐ 5 White ☐ 6 Mixed ☐ 8 Other ☐ 9 Unknown

11. Are you of Hispanic origin? ☐ 1 Yes ☐ 2 No ☐ 9 Unknown

12. Are you comfortable speaking in English for this interview? ☐ 1 Yes ☐ 2 No
    If the person answers no, go to 12a.

12.a What is your preferred language?_____
(Interviewer, stop and arrange to call back with an interviewer in the preferred language if necessary.)

13. What is your birth date? __ __/__ __/__ __ (mm/dd/yy) Estimated? ☐ Y ☐ N Basis _____

Now I would like to ask you some questions about when you were exposed to pesticides.

Event Information Screen, Application/Release Event Narrative

14. Can you briefly describe the events leading up to your pesticide exposure?
_____
_____
_____
_____

Event Information Screen, Event Summary—Application Information

15. Where did the application (*or event such as a spill, transport accident, or fire*) that was associated with your exposure take place? (*Interviewer, enter the code from the list below. Do not read options.*)

☐☐

01	Farm (excluding, nursery, livestock, forest)	32	Farm product warehousing and storage
02	Nursery	33	Food manufacturing
03	Forest	39	Other manufacturing facility/industrial facility/warehouse facility
04	Livestock and other animal specialty production facility	40	Office/business (nonretail, nonindustrial)
05	Greenhouse	41	Retail establishment
09	Other agricultural processing facility	42	Service establishment
10	Single family home	43	Pet care services and veterinary facilities
11	Mobile home	50	Road/rail
12	Multiunit housing (apartments, multiplexes)	51	Road, rail, or utility right-of-way
13	Labor housing	52	Park
20	Residential institution (dorms, shelters)	54	Private vehicle
21	School	55	Public transportation vehicle
22	Day care facility (including in private residence)	59	Other
23	Prison	60	Emergency response vehicle
24	Hospital	70	More than one site
29	Other institution	98	Not applicable
30	Pesticide manufacturing/formulation facility	99	Unknown

Rev. 7/1/04

APPENDIX C ■ SECTION 2
PESTICIDE EXPOSURE QUESTIONNAIRE

Pesticide Illness and Injury Surveillance Questionnaire

Case ID __ __ __ __ __ __ __ __ __

Event ID __ __ __ __ __ __ __ __ __

16a. What was the intended target for the pesticide?

(Interviewer, mark only one from the list. Do not read options.)

- ☐ (060) Aquatic (pond, stream, lake, irrigation canal)
- ☐ (800) Bait for rodent, bird, or predator
- ☐ (200) Beverage crops
- ☐ (041) Building structure (including crack and crevice treatment.)
- ☐ (042) Building surface
- ☐ (043) Building space treatment
- ☐ (530) Cereal grain crops (e.g., barley, corn, wheat, rice)
- ☐ (650) Crops that cross categories 90–600 (general farming)
- ☐ (801) **Community-wide application target** *(go to 16b below)*
- ☐ (501) Fiber crops (e.g., cotton)
- ☐ (300) Flavoring and spice crops
- ☐ (510) Forage, fodder hay, silage grasses, silage legumes, and related crops
- ☐ (020) Forest trees and forest lands
- ☐ (100) Fruit crops
  - ☐ (110) Tree fruits
    - ☐ (111) Citrus fruits (e.g., grapefruit, kumquat, lemon, oranges)
    - ☐ (113) Pome fruits (e.g., apples, pears, quince, Japanese plum)
    - ☐ (101) Small fruits (e.g., berries, currants, grapes)
    - ☐ (114) Stone fruits (e.g., apricots, cherries, dates, mangoes, olives)
  - ☐ (120) Subtropical/other fruits (e.g., avocado, banana, coconuts)
  - ☐ (112) Tree nuts (e.g., almonds, hazelnuts, pecans)
- ☐ (500) Grains, grasses, and fiber crops
- ☐ (700) Human
  - ☐ (701) Human—skin/hair
  - ☐ (702) Human—clothing
  - ☐ (703) Human—skin/hair and clothing
- ☐ (010) Landscape/ornamental
- ☐ (550) Miscellaneous field crops
- ☐ (600) Oil crops
- ☐ (850) Other (e.g., mixed crop and noncrop, mammal feeding and nesting areas, boats and docks)
- ☐ (601) Seed treatment (application to seeds)
- ☐ (070) Soil
- ☐ (540) Sugar crops (e.g., sugar cane, sorghum)
- ☐ (050) Undesired plant (the plant is the target pest)
- ☐ (400) Vegetable crops
  - ☐ (410) Curcubit vegetables (e.g., cucumbers)
  - ☐ (420) Fruiting vegetables (e.g., cantaloupe, melon, squash)
  - ☐ (430) Leafy vegetables (e.g., cabbage, celery, endive, lettuce)
  - ☐ (460) Other vegetables (e.g., broccoli, cauliflower, eggplant)
  - ☐ (440) Root and tuber vegetables (e.g, beets, carrots, onions)
  - ☐ (450) Seed and pod vegetables (e.g., beans, chick-peas, lentils, peanuts, peas, soybeans, sweet corn)
- ☐ (032) Veterinary/domestic animal
- ☐ (031) Veterinary/livestock
- ☐ (080) Wood product (e.g., utility poles, decking, fencing, boardwalk, railroad ties, bulwarks, pilings)
- ☐ (998) Not applicable, application not involved
- ☐ (999) Unknown

If 16a is coded as community-wide application target 801, complete items 16b and 16c.

16b. What was the purpose of the community-wide application?

- ☐ 1 Agricultural pest eradication
- ☐ 2 Public health pest control or eradication
- ☐ 8 Not Applicable
- ☐ 9 Unknown

Rev. 7/1/04

APPENDIX C ■ SECTION 2
PESTICIDE EXPOSURE QUESTIONNAIRE

Pesticide Illness and Injury Surveillance Questionnaire

Case ID _ _ _ _ _ _ _ _

Event ID _ _ _ _ _ _ _ _

16c. What was the specific target of the community-wide application? ☐ ☐ ☐

001	Mosquito (no disease specified)	103	Japanese beetle
002	West Nile virus	104	Imported fire ant (red or black)
003	St. Louis encephalitis	105	Asian longhorn beetle
004	Eastern equine encephalitis	106	Emerald ash borer
005	Western equine encephalitis	107	Grain fungal diseases (e.g., black stem rust)
006	La Crosse encephalitis	108	Grasshopper/Mormon cricket
007	Dengue fever	888	Default if State chooses not to code this variable
100	Boll weevil	996	Multiple pests
101	Gypsy moth (Asian or European)	998	Not applicable (APPTARGT not = 801)
102	Fruit fly (Mediterranean, Mexican, Oriental, olive, etc.)	999	Unknown

17. What type of equipment was used in this application? (*Interviewer, mark only one from the list below. Do not read options.*) ☐ ☐

01	Aerial application equipment	10	Trigger pump/compressed air
02	Chemigation	11	Ground sprayer
03	Pressurized can	12	Manual placement
04	Aerosol generator/fogger	13	Dip tank or tray
05	Soil injector	14	More than one type of equipment
06	High-pressure fumigator	15	Other
07	Hand-held granular/dust application	98	Not applicable
08	Spray line, hand-held	99	Unknown
09	Sprayer, backpack		

Event Information Screen, Location

18. What is the address where the **event** occurred that is associated with this exposure? This address is the site of the pesticide application, spill, or release (that is, field, orchard, business, institution, residence, or roadway). (*For locations without specific addresses, include closest crossroad and distances. This may differ from a person's location at the time of exposure. For example, the exposed person might be located at a school and the actual event is a fire at a nearby pesticide storage facility. The event location is the pesticide storage facility.*)

Address line 1 _____

Address line 2 _____

City _____

State __ __   ZIP __ __ __ __ __   Latitude _____

County name _____   FIPS __ __ __   Longitude _____

APPENDIX C ■ SECTION 2
PESTICIDE EXPOSURE QUESTIONNAIRE

Pesticide Illness and Injury Surveillance Questionnaire

Case ID __ __ __ __ __ __ __ __

Event ID __ __ __ __ __ __ __ __

Next, I am going to ask you some questions about the pesticide products you were exposed to and how you were exposed.

Event Information Screen, Pesticide Products

Interviewer, complete the information below (complete as much of the information as possible for each chemical) by asking the following questions:

19. What is the name of the chemical that you were exposed to? If you were exposed to more than one chemical, please tell me the name of each one. (*Interviewer, record all information available including manufacturer and any modifiers on label, e.g., spray, dust, 4E.*)

EPA registration number/distributor number	Name	Form*	Poisoning attribution†
a. __ __ __ __ __ __ __ - __ __ __ __ __ __ / __ __ __ __ __ __	_____ _____ _____	_____	☐
b. __ __ __ __ __ __ __ - __ __ __ __ __ __ / __ __ __ __ __ __	_____ _____ _____	_____	☐
c. __ __ __ __ __ __ __ - __ __ __ __ __ __ / __ __ __ __ __ __	_____ _____ _____	_____	☐
d. __ __ __ __ __ __ __ - __ __ __ __ __ __ / __ __ __ __ __ __	_____ _____ _____	_____	☐
e. __ __ __ __ __ __ __ - __ __ __ __ __ __ / __ __ __ __ __ __	_____ _____ _____	_____	☐

*See form codes on next page.
†Check box if product is thought to have contributed to illness. Complete this column at time of case closure.

Interviewer, if the EPA Registration Number is unknown, but the identity of active ingredient(s) is known, enter the most detailed product name available above and the active ingredients below. Information about products where only form and functional class are known, and information about carriers and inerts can be section labeled **Other Pesticide Information** (Item 21, page 7). If the EPA Registration Number is entered in the known, complete the EPA Registration Number and Product name, then skip to Item 22 **Chemical Agent Comments, page 7.**

Rev. 7/1/04

APPENDIX C ■ SECTION 2
PESTICIDE EXPOSURE QUESTIONNAIRE

Pesticide Illness and Injury Surveillance Questionnaire

Case ID _ _ _ _ _ _ _

Event ID _ _ _ _ _ _ _

Event Information Screen, Pesticide Product—Active Ingredients and Other Sources

20. Active Ingredient. *If product name is unknown but active ingredient is known, enter active ingredient here. (Code is auto entered in SPIDER; record only if using lookup file for entry into a nonautomated system.)*

Active ingredient code	Name	Per-centage	Form*	Chemical class*	Functional class*	Poisoning attribution[†]
a.						☐
b.						☐
c.						☐
d.						☐

*Indicate the product form, chemical, and functional class from the tables below.

† Check box if product is thought to have contributed to illness. Complete this column at time of case closure.

NIOSH form codes			
01	Dust/powder (not pressurized)	10	Flowable concentrate
02	Granular/flake	11	Pressurized liquid/spray/fogger
03	Pellet/tablet/cake/briquette	12	Ready-to-use liquid/solution
04	Wettable powder/dust	13	Other liquid formulation
05	Impregnated material (ant/plant stakes, animal collars, water filters)	14	Pressurized gas/fumigant
		15	Paint/liquid coating
06	Other dry formulation	16	Other
07	Microencapsulated	17	Soluble powder
08	Emulsifiable concentrate	18	Liquid concentrate
09	Soluble concentrate	99	Unknown

Chemical class codes	Functional class codes
01 Organochlorine compound	01 Insecticide (excluding solely IGR and fumigants)
02 Organophosphorous compound	02 Insect growth regulator (IGR)
03 N-methyl carbamates	03 Herbicide/algicide
04 Pyrethrin	04 Fungicide
05 Pyrethroid	05 Fumigant
06 Dipyridyl compound	06 Rodenticide
07 Chlorophenoxy compound	07 Disinfectant/broad spectrum for water sanitation
08 Triazines	08 Insect repellent
09 Carbamates (non-AChE inhibitors)	09 Antifouling agent (marine paints)
10 Organo-metallic compound	10 Insecticide and herbicide (01 & 03)
11 Inorganic compounds	11 Insecticide and fungicide (01 & 04)
12 Coumarins	12 Insecticide and herbicide and fungicide (01, 03, & 04)
13 Indandiones	13 Insecticide and other (01 & 96)
14 Convulsants	14 Herbicide and fungicide (03 & 04)
15 Microbial	96 Other (includes biological controls, plant growth regulators, antibiotics, etc.)
16 Dithiocarbamates	
95 Unidentified cholinesterase inhibitor	97 Multiple (product is classified as multiple classes which do not fit in any of the codes specified in codes 10–14)
97 Multiple (PC Code indicates a code for a combination of active ingredients that cross chemical classes)	99 Unknown
99 Unknown	

Rev. 7/1/04

APPENDIX C ■ SECTION 2
PESTICIDE EXPOSURE QUESTIONNAIRE

Pesticide Illness and Injury Surveillance Questionnaire

Case ID _ _ _ _ _ _ _ _

Event ID _ _ _ _ _ _ _ _

21. Other Pesticide Information. If neither product nor active ingredient is known.
Enter a description of the other pesticide, e.g., "Some kind of spray from a highway truck"; "unspecified Black Flag wasp spray"; "unlabeled spray can." This area can also be used to record information about carriers and inerts at the State level.

Other ID	Description of other source	Chemical class*	Form*	Functional class*	Poisoning attribution[†]
a.					☐
b.					☐
c.					☐

* Indicate the product form, chemical, and functional class from the tables above.
[†]*Check box if product is thought to have contributed to illness. Complete this column at time of case closure.*

22. Event Information, Comments—Chemical Agent Comments
(Note additional information about pesticide products and adjuvants.)

_____
_____
_____
_____
_____

Interviewer, complete this section after completing interview. All ID numbers are assigned by SPIDER upon data entry. Data entry clerk should enter ID numbers onto form.

Event ID _ _ _ _ _ _ _   Event descriptor _____
                                              (Maximum 30 character name for event)

23. Was the pesticide applied by a licensed applicator? (*Indicate the level of applicator supervision. This may require interviewing affected person, employer, or contract applicator to determine response.*) ☐ (1)Licensed applicator ☐ (2)Licensed trainee, direct supervision ☐ (3)Unlicensed, intermittent supervision ☐ (4)Unlicensed ☐ (8)Not applicable ☐ (9)Unknown	24. Is there evidence indicating that the label directions were <u>not</u> followed? ☐ (1)Yes, there is evidence that label directions were not followed. ☐ (2)No, no evidence of label directions not being followed. ☐ (8)Not applicable. ☐ (9)Unknown.

Rev. 7/1/04   -7-

APPENDIX C ■ SECTION 2
PESTICIDE EXPOSURE QUESTIONNAIRE

Pesticide Illness and Injury Surveillance Questionnaire

Case ID __ __ __ __ __ __ __

Event ID __ __ __ __ __ __ __

Exposure Information Screen, Incident Information

Incident report information. *Interviewer, complete Items 25–28 prior to interview. All ID numbers assigned by SPIDER on data entry. Data entry clerk, enter onto form here.*

Exposure ID _____	25. Report date __ __ / __ __ / __ __ __ __	
Case ID __ __ __ __ __ __		Event ID __ __ __ __ __ __
26. Report source 1 ☐ ☐ ☐	27. Report source 2 ☐ ☐ ☐	28. Report source 3 ☐ ☐ ☐

Use codes for sources below. Note that an additional character can be added for State-specific codes under each category, (e.g., listing specific poison control centers in the State by using codes 020-029 or 02A – 02Z).

Source Code	Description
01	Physician report
02	Poison control center
03	Other health care provider report (including emergency room or hospital report)
04	Laboratory report
05	Death certificate or medical examiner's report
06	Report or referral from governmental agency
07	Obituary/news report
08	Ascertainment through Worker's Compensation
09	Self-report
10	Co-worker report
11	Friend or relative report
12	Identified during site visit
13	Worker representative (e.g., union, lawyer/legal services/other advocate)
14	Medical record review (clinic or hospital record review performed by surveillance staff)
97	State Department of Health
98	Other (not captured in any code category listed)
99	Unknown

29. Where were you when the exposure took place? *(Interviewer, enter from the list below. Do not read options, but base entry upon verbal response.)* ☐ ☐ ☐

01	Farm (excluding, nursery, livestock, forest)	32	Farm product warehousing and storage
02	Nursery	33	Food manufacturing
03	Forest	39	Other manufacturing facility/industrial facility/warehouse facility
04	Livestock and other animal specialty production facility	40	Office/business (nonretail, nonindustrial)
05	Greenhouse	41	Retail establishment
09	Other agricultural processing facility	42	Service establishment
10	Single family home	43	Pet care services and veterinary facilities
11	Mobile home	50	Road/rail
12	Multiunit housing (apartments, multiplexes)	51	Road, rail, or utility right-of-way
13	Labor housing	52	Park
20	Residential institution (dorms, shelters)	54	Private vehicle
21	School	55	Public transportation vehicle
22	Day care facility (including in private residence)	59	Other
23	Prison	60	Emergency response vehicle
24	Hospital	70	More than one site
29	Other institution	98	Not applicable
30	Pesticide manufacturing/formulation facility	99	Unknown

Rev. 7/1/04

APPENDIX C ■ SECTION 2
PESTICIDE EXPOSURE QUESTIONNAIRE

Pesticide Illness and Injury Surveillance Questionnaire

Case ID __ __ __ __ __ __ __ __
Event ID __ __ __ __ __ __ __ __

30. What is the address for the location where you were exposed? (*This may be the same as the case address or the event address.*)

Address 1	_____
Address 2	_____
City	_____
State	__ __ ZIP __ __ __ __ __
Latitude	_____ Longitude _____
County name	_____ FIPS __ __ __

31. What were you doing when you were exposed? (*Interviewer, determine appropriate code for the response; do not read from the list of coded options. Check one only.*)

- ☐ (01) Applying pesticide
- ☐ (02) Mixing/loading pesticide
- ☐ (03) Transport or disposal of pesticide
- ☐ (04) Repair or maintenance of pesticide application equipment
- ☐ (05) Any combination of activities 01–04
- ☐ (06) Involved in manufacture or formulation of pesticide
- ☐ (07) Emergency response
- ☐ (08) Routine work activities not involved with pesticide application (includes exposure to field residue)
- ☐ (09) Routine indoor living activities not involved with pesticide application
- ☐ (10) Routine outdoor living activities not involved with pesticide application
- ☐ (98) Not applicable
- ☐ (99) Unknown

32. Were other people possibly exposed?   ☐ (1) Yes   ☐ (2) No   ☐ (9) Unknown

> **If yes, continue with 32a and 32b.**
>
> 32a. How many _____
>
> 32b. Did any seek medical care?   ☐ (1) Yes   ☐ (2) No   ☐ (9) Unknown

Use a separate sheet of paper to record names and contact information, if appropriate.

Rev. 7/1/04

APPENDIX C ■ SECTION 2
PESTICIDE EXPOSURE QUESTIONNAIRE

Pesticide Illness and Injury Surveillance Questionnaire

Case ID __ __ __ __ __ __ __
Event ID __ __ __ __ __ __ __

33. Please describe the exposure to me, especially anything we haven't yet discussed. I may ask you some more detailed questions about what you describe as we proceed with the interview.

_____
_____
_____
_____
_____
_____
_____

34. When you were exposed to the pesticide, did you seek any type of medical care?
    ☐ (2) No   ☐ (1) Yes
    ☐ (9) Unknown

    Go to item 44a on page 11.   **Complete items 35–43.**

35. Where did you receive your initial medical care after the exposure?
    ☐ (1) Physician office/clinic visit   ☐ (5) No medical care sought
    ☐ (2) Emergency room   ☐ (6) Other
    ☐ (3) Hospital admission   ☐ (9) Unknown
    ☐ (4) Advice from the poison control center

36. When did you first receive medical care? __ __ / __ __ / __ __
    (mm/dd/yy)

37. What is the name of the health care professional (HCP) you saw?

    Last name _____
    First name _____

38. What is their address?

	Chart location	**Work location**
Address 1	_____	_____
Address 2	_____	_____
City	_____	_____
State	__ __	__ __
Zip	__ __ __ __ __	__ __ __ __ __
Phone	(__ __ __) __ __ __ - __ __ __ __	(__ __ __) __ __ __ - __ __ __ __

Rev. 7/1/04

APPENDIX C ■ SECTION 2
PESTICIDE EXPOSURE QUESTIONNAIRE

Pesticide Illness and Injury Surveillance Questionnaire

Case ID ___ ___ ___ ___ ___ ___ ___ ___

Event ID ___ ___ ___ ___ ___ ___ ___ ___

39. Did you have a test for pesticides in your blood or urine?

☐ (2) No   ☐ (1) Yes

☐ (9) Unknown   *Interviewer, obtain medical record to complete tables on pages 18-19.*

Go to item 41.

40. Did you have a cholinesterase test, which requires that blood be drawn?

☐ (2) No   ☐ (1) Yes   ☐ (9) Unknown

If yes, complete tables on pages 18-19 from the medical record.

41. Were you admitted to the hospital due to the pesticide exposure?   ☐ (2) No   ☐ (1) Yes   ☐ (9) Unknown

If No or Unknown, go to item 44a.

42. Facility where hospitalized _____

_____

Address_____

_____

Treating physician_____

43. How many days did you stay in the hospital? ___ ___ ___

(Enter code number of days: 997 if ≥ 996 days, 998=NA, not hospitalized 999=Unknown or 999 if unknown.)

44a. Did you spend one or more days away from work due to the pesticide exposure?

☐ (1) Yes   ☐ (2) No

                                          ☐ (9) Unknown

44b. How many days were you away from work? _____

Go to item 44c

Rev. 7/1/04

-11-

130

APPENDIX C ■ SECTION 2
PESTICIDE EXPOSURE QUESTIONNAIRE

Pesticide Illness and Injury Surveillance Questionnaire

Case ID _ _ _ _ _ _ _ _
Event ID _ _ _ _ _ _ _ _

44c. If not employed, did you spend one or more days away from school or regular activities?

☐ (1) Yes ☐ (2) No ☐ (9) Unknown

44d. How many days were you away from school or regular activities? _____

45. Do you have any of the following medical conditions that were not due to this exposure?

Condition (Check all that apply.)	Describe	(Interviewer, complete after interview.) Medical history (from HCP interview or medical records)	Final code*
___a. Pregnancy			
___b. Asthma			
___c. Allergies			
___d. Multiple chemical sensitivity (acquired chemical intolerance)			
___e. Any other medical condition you are seeing a doctor for			

*Use the following codes for Final code after completion of interview and medical record review
1=Doctor reported   2=Exposed person reported   3=Both doctor and person reported
4=Condition was absent 5=Not reported   9=Unknown

Interviewer, was response to question 31 on page 9 in the range of 01-08?
YES—proceed to question 46.
NO—skip to question 63 on page 14.

Exposure Information Screen, Occupational Information, PPE Use

46. Did the pesticide exposure occur while you were working?

☐ (1) Yes   ☐ (3) No
☐ (2) Possibly   ☐ (4) Unknown
                             ☐ (5) Not Applicable
                                 ↓
                             Go to Item 54 on page 13.

Now I am going to ask you some questions about your employer **at the time you were exposed to pesticides.** *(If necessary, reassure the subject that you are not associated with OSHA.)*

Rev. 7/1/04

APPENDIX C ■ SECTION 2
PESTICIDE EXPOSURE QUESTIONNAIRE

Pesticide Illness and Injury Surveillance Questionnaire

Case ID _ _ _ _ _ _ _ _

Event ID _ _ _ _ _ _ _ _

47. What is your employer's name? _____

48. What is your employer's address (or your work address if self-employed)?

   Street Address _____

   City _____ State __ __ Zip code __ __ __ __ __

49. What was your occupation/job title when you were injured/exposed? _____

50. What type of work was being done at your place of employment at time of injury/exposure?

_____

> *Is this individual an agricultural worker or pesticide handler?*
> *YES—Complete item 51a.*
> *NO—Proceed to item 54.*

51a. Did this incident involve entering a treated area (including field or greenhouse)?

　　☐ (2) No　　☐ (1) Yes　　☐ (9) Unknown

If yes, ask 51b. Did your employer/crew leader tell you how soon you could go into the area after it was treated?

　　☐ (2) No　　☐ (1) Yes　　☐ (9) Unknown

To be completed by the interviewer after interview.

52. *Bureau of Census code for occupation of exposed worker* __ __ __
　　　　　　　　　　　　　　　　　　　　　　　　　　　　Occ code

53. *Bureau of Census code for industry or North American Industry Classification System (NAICS)*
　　__ __ __ *or* __ __ __ *(Note that Census codes are preferred.)*
　　Indcic　　　　Indsic

54. Were you wearing any personal protective equipment?

　　　　☐ (2) No　　☐ (1) Yes
　　　　☐ (9) Unknown

　　Go to item 63　　　　　*Go to items 55-61*

Rev. 7/1/04

APPENDIX C ■ SECTION 2
PESTICIDE EXPOSURE QUESTIONNAIRE

Pesticide Illness and Injury Surveillance Questionnaire

Case ID __ __ __ __ __ __ __ __ __

Event ID __ __ __ __ __ __ __ __ __

55. Were you wearing
   a. a supplied air respirator?
      ☐ (1) Yes   ☐ (2) No   ☐ (9) Unknown

   b. half/full face, PAPR?
      ☐ (1) Yes   ☐ (2) No   ☐ (9) Unknown

56. rubber/chemically resistant boots?
      ☐ (1) Yes   ☐ (2) No   ☐ (9) Unknown

57. cloth or leather gloves?
      ☐ (1) Yes   ☐ (2) No   ☐ (9) Unknown

58. rubber or synthetic gloves?
      ☐ (1) Yes   ☐ (2) No   ☐ (9) Unknown

59. chemical goggles/face shield?
      ☐ (1) Yes   ☐ (2) No   ☐ (9) Unknown

60. chemically resistant clothing? (rubber apron, tyvek, rain gear)
      ☐ (1) Yes   ☐ (2) No   ☐ (9) Unknown

61. Were you using engineering controls? (e.g., closed mixing/loading system)
      ☐ (1) Yes   ☐ (2) No   ☐ (9) Unknown

62. *Interviewer, complete after interview. Indicate the level of PPE used and required for this individual according to the product label.*

☐ (1) Used (all or some of PPE required)
☐ (2) Used (not required)
☐ (3) Used (unknown requirements)
☐ (4) Not used (some PPE required)
☐ (5) Not used (unknown requirements)
☐ (6) Not used (not required)
☐ (8) Not applicable
☐ (9) Unknown

63. What was the date and approximate time that your exposure to the pesticide(s) first started?
   Date* ___/___/___ Time ___:___ Use 24-hour clock
        (mm/dd/yy)

At least one of the following dates must be entered: first exposure, symptom onset, or laboratory test (see pages 18-19).

Rev. 7/1/04

APPENDIX C ■ SECTION 2
PESTICIDE EXPOSURE QUESTIONNAIRE

Pesticide Illness and Injury Surveillance Questionnaire

Case ID __ __ __ __ __ __ __ __
Event ID __ __ __ __ __ __ __ __

64. What was the date and approximate time that you first started to experience symptoms?
Date ___/___/___ Time ___:___ Use 24-hour clock
(mm/dd/yy)

65. What was the date and approximate time that your exposure to the pesticide(s) ended?
Date ___/___/___ Time ___:___ Use 24-hour clock
(mm/dd/yy)

Signs/Symptoms

66. Next I'd like you to describe your symptoms. (*Interviewer, fill in "Doctor reported" column based on medical records or HCP interview. Final code column should be completed prior to data entry. Codes are listed following the table on page 17.*)
Check all signs or symptoms described or stated as absent (items in italics should be taken from HCP interview or medical record only).

System	Sign/symptom	Patient reported	Doctor reported	Absent	Final code
General	*Acidosis*				
	Alkalosis				
	Fatigue/malaise				
	Fever				
	Increased anion gap				
	Other_____				
Cardiovascular	*Bradycardia*				
	Cardiac arrest				
	Chest pain				
	Conduction disturbance				
	Hypertension				
	Hypotension				
	Palpitations				
	Tachycardia				
	Other_____				
Renal	Frequent urination				
	Hematuria				
	Oliguria/anuria				
	Proteinuria				
	Other_____				
Neurological	Altered taste				
	Anxiety/hyperactivity/irritability				
	Ataxia/trouble walking				
	Blurred vision				

Rev. 7/1/04

APPENDIX C ■ SECTION 2
PESTICIDE EXPOSURE QUESTIONNAIRE

Pesticide Illness and Injury Surveillance Questionnaire

Case ID __ __ __ __ __ __ __
Event ID __ __ __ __ __ __ __ __

System	Sign/symptom	Patient reported	Doctor reported	Absent	Final code
Neurological (continued)	Coma				
	Confusion				
	Diaphoresis (profuse sweating)				
	Dizziness				
	Fainting				
	Fasciculations				
	Headache				
	Memory loss				
	Muscle pain				
	Muscle rigidity				
	Muscle weakness				
	Paralysis				
	Paresthesias/tingling or numbness				
	Peripheral neuropathy				
	Salivation				
	Seizure				
	Slurred speech				
	Other_____				
Gastrointestinal	Anorexia (loss of appetite)				
	Constipation				
	Diarrhea				
	GI bleeding (blood in stool or vomit)				
	Nausea				
	Pain				
	Vomiting				
	Other_____				
Eye	*Burns*				
	Conjunctivitis (diagnosis)				
	Corneal abrasion				
	Miosis				
	Mydriasis				
	Pain/irritation/inflammation				
	Tearing/*lacrimation*				
	Other_____				

Rev. 7/1/04

APPENDIX C ■ SECTION 2
PESTICIDE EXPOSURE QUESTIONNAIRE

Pesticide Illness and Injury Surveillance Questionnaire

Case ID _ _ _ _ _ _ _ _
Event ID _ _ _ _ _ _ _ _

System	Sign/symptom	Patient reported	Doctor reported	Absent	Final code
Dermal	Blisters/*bullae*				
	Burns				
	Edema/swelling				
	Hives				
	Pain				
	Pruritis (itching)				
	Pattern* of rash or lesions				
	Rash				
	Redness				
	Other_____				
Respiratory	*Asthma (diagnosis of)*				
	Cough				
	Cyanosis				
	Depression				
	Dyspnea				
	LR Irritation				
	Pleural pain				
	Pulmonary edema				
	Tachypnea				
	UR irritation				
	Wheezing				
	Other_____				

*Coding for pattern of dermal lesions
1=Corresponds well with physical pattern of exposure
2=Discrete patches of lesions do not correspond with the pattern of exposure
3=Generalized distribution of lesions on the body
4=Absent
9=Unknown

Complete final code column prior to data entry.

Final Code for All Fields 1=Doctor reported   2=Exposed person reported   3=Both Dr. and person reported
9=Unknown

Rev. 7/1/04
- 17 -

APPENDIX C ■ SECTION 2
PESTICIDE EXPOSURE QUESTIONNAIRE

Pesticide Illness and Injury Surveillance Questionnaire

Case ID ___ ___ ___ ___ ___ ___ ___ ___

Event ID ___ ___ ___ ___ ___ ___ ___ ___

Exposure Information Screen, Narrative

Health Comments_____

_____

_____

_____

Ending Statement

This concludes the interview. The information you have given us is very important. We appreciate your willingness to take time to answer all of our questions. Do you have any questions at this time? *(Interviewer, provide the caller with your name and phone number and information about any additional contacts or actions that will result from the interview.)*

The following section should be completed after reviewing medical records or interviewing the attending HCP. Also, go to pages 15-17 to update signs and symptoms based on the medical records/HCP interview, if this section is not pertinent. Make sure diagnosis, outcome, and any notes are entered on page 19.

Exposure Information Screen, Medical Staff

Medical ID	___ ___ ___ ___ ___ ___ ___ ___ ___ ___ ___ ___ ___ ___ ___ ___ ___ ___
Enter one Medical ID for each medical person involved in the case. In SPIDER, use F2 to select from pick list. If not on pick list, see Item 37 on Page 10 and enter on Medical Staff screen the full medical staff information.	

Interviewer, complete this section and shaded columns on the table for Item 45 on page 12, based on interview with HCP or review of medical records.

Non-Cholinesterase Chemical-Specific Biological Test for Pesticides or Metabolites

Were any non-cholinesterase biological tests done for pesticides in blood, urine, or hair?	☐ Yes ☐ No ☐ Unknown *If yes, complete table if part of State protocol*			
	Test 1		**Test 2**	
Test type				
Sample date	___ ___ / ___ ___ / ___ ___		___ ___ / ___ ___ / ___ ___	
Numeric result				
Analysis result	☐ Abnormal ☐ Not applicable	☐ Normal ☐ Unknown	☐ Abnormal ☐ Not applicable	☐ Normal ☐ Unknown

Rev. 7/1/04

APPENDIX C ■ SECTION 2
PESTICIDE EXPOSURE QUESTIONNAIRE

Pesticide Illness and Injury Surveillance Questionnaire

Case ID __ __ __ __ __ __ __ __
Event ID __ __ __ __ __ __ __ __

Was a cholinesterase test(s) performed? *(Ask only if exposure involved an organophosphate or N-methyl carbamate pesticide.)* ☐ (1) Yes   ☐ (2) No   ☐ (9) Unknown

(If yes, complete table of results below.)

Coding Guidance for Completing Table of Results Below
Option 1: Detailed version—complete all.
Option 2: Required minimum—enter a single response for *Test Type* and *Result Type*, the only required fields.

PFI	Lab code from lab pick list or enter lab name						
Test type codes	1=RBC 2=Plasma 3=Both RBC and Plasma 4=Not done 5=Either RBC or Plasma 8=Not applicable 9=Unknown						
Result type codes	1=Abnormal compared to lab 2=Abnormal compared to baseline 3=Normal compared to lab 4=Normal compared to baseline 7=Bad specimen 8=Not applicable 9=Unknown						
PFI	Lab name	Test type	Test date	Numeric result	Result type	Lab low	Lab high
			__/__/__				
			__/__/__				
			__/__/__				
			__/__/__				
			__/__/__				

[ ] **Diagnosis made by HCP**	**Outcome**
Diagnosis _____ _____ ICD-9 __ __ __ . __ __ Summary _____ _____	☐ (1) Fatal, pesticide-related ☐ (2) Fatal, not pesticide-related ☐ (3) Fatal, relation unknown ☐ (8) Not applicable (not fatal)

Notes _____
_____
_____
_____

Rev. 7/1/04

APPENDIX C ■ SECTION 2
PESTICIDE EXPOSURE QUESTIONNAIRE

Pesticide Illness and Injury Surveillance Questionnaire

Case ID __ __ __ __ __ __ __ __

Event ID __ __ __ __ __ __ __ __

Interviewer, review the form for completeness and complete the following sections:

Pages 4 – 7   Check to determine if other interviews are needed to complete questions 18–22.
Pages 5 – 7   Any necessary chemical product coding.
Page 7   Event descriptor, items 23 and 24.
Page 12   Item 45 enter shaded columns: pre-existing conditions from medical professional interview or medical record review and complete final code for medical conditions.
Page 13   Items 52 and 53.
Page 14   Item 62.
Pages 15 –17   Make sure all signs and symptoms are entered from medical professional interview or medical record review and complete final code column.

Indicate your assessment of how the individual came into contact with the pesticide. *(Check all that apply; bolding indicates variable label in SPIDER.)*

☐ **Drift**
☐ Direct **spray**
☐ **Indoor air** contamination
☐ Contact with treated **surface** (plant material, carpets, treated animal)
☐ Direct **contact** (spill, leaking container or equipment, floodwaters, emergency response)
☐ **Other**
☐ **Unknown**

Indicate the route(s) of exposure. *(Check all that apply.)*

☐ Dermal   ☐ Injection
☐ Inhalation   ☐ Ocular
☐ Ingestion   ☐ Unknown

Indicate if the exposure was intentional.

☐ 1=Yes, suspected intentional   ☐ 2=No, unintentional   ☐ 9=Unknown

Rev. 7/1/04

APPENDIX C ■ SECTION 2
PESTICIDE EXPOSURE QUESTIONNAIRE

Pesticide Illness and Injury Surveillance Questionnaire

Case ID __ __ __ __ __ __ __ __

Event ID __ __ __ __ __ __ __ __

The remaining sections are to be completed by the interviewer following standard case classification procedures

Severity ☐
    1 = Fatal   2 = High   3 = Moderate   4 = Low   8 = Evaluated, Not applicable

A. Documentation of Exposure ☐ ☐
(Put a number in the first box and letter in the second box if appropriate.)

1 - Confirmed by   a-envir/bio testing   b-professional observation   c-biological evidence
                   d-eye/derm signs   e. 2+ findings by medical staff

2 - Reported by   a-case   b-witness   c-application records
                 d-nonprofessional observation   e-other

3 - Strong evidence of no exposure

4 - Insufficient data

B. Documentation of Health Effect ☐
    1 - 2+ findings by medical staff
    2 - 2+ abnormal symptoms
    3 - No post exposure findings
    4 - Insufficient Data

C. Evaluation of Causal Relationship ☐ ☐
(Put a number in first box and letter in second box if first box is 1.)
    1 - Fits known toxicology
        a-characteristic (Appendix 2 of case classification) and temporal relationship is plausible
        b-consistent with literature and known toxicology
    2 - Inconsistent with known toxicology
    3 - Definitely ruled out (evidence of non-pesticide causal agent)
    4 - Insufficient toxicologic information available

NIOSH Classification ☐   **Alternate Classification** ☐

Classification categories   1=Definite   5=Unlikely
                                    2=Probable   6=Insufficient Information
                                    3=Possible   7=Exposed/Asymptomatic
                                    4=Suspicious   8=Unrelated

Exposure Information Screen–Poisoning Attribution

Return to pages 5 through 7 to determine if illness is attributable to products, active ingredients, or substances listed there.

Rev. 7/1/04

APPENDIX C ■ SECTION 3
DATA COLLECTION FORM

C.3 PESTICIDE ILLNESS AND INJURY SURVEILLANCE DATA COLLECTION FORM

INSTRUCTIONS

This is a sample data collection form for use by an acute pesticide-related illness and injury surveillance program. This form is for States that choose not to use a standard questionnaire but collect data via an open-ended interview. The form includes fields that satisfy data requirements for all of the core variables needed by the National Institute for Occupational Safety and Health (NIOSH), Centers for Disease Control and Prevention (CDC). Fields needed for administrative report management at the State level, as well as optional suggested variables are also included. Optional items are indicated on the form by framing with a dotted-line border. The order of the fields is designed to provide ease of data collection as well as data entry using the SENSOR Pesticide Incident Data Entry and Reporting (SPIDER) data management software. Shading indicates items that are to be completed by the interviewer and not asked during the actual exposed individual or attending health care professional (HCP) interview. The form contains fields for information that may be collected from the exposed individual, and additional medical and pesticide product information collected from record reviews or additional interviews. States will need to customize this data collection form for their specific needs.

APPENDIX C ■ SECTION 3
FORM CONTAINS CONFIDENTIAL INFORMATION

PESTICIDE ILLNESS AND INJURY SURVEILLANCE DATA COLLECTION FORM

Case ID ___ ___ ___ ___ ___ ___ ___   Event ID ___ ___ ___ ___ ___ ___ ___ ___ ___

Case Information

Social Security Number ___ ___ ___ – ___ ___ – ___ ___ ___ ___

Name

_____ _____ _____
Last   First   Initial

DOB ___/___/_____ Estimated? ☐Y ☐N Basis_____
         MM   DD   YYYY

DOD ___/___/_____
         MM   DD   YYYY

Current residence information

Address Line 1 _____

Address Line 2 _____

City _____

State ___ ___   Zip ___ ___ ___ ___ ___   County _____

Phone (___ ___ ___) ___ ___ ___ – ___ ___ ___ ___

Residence at time of exposure if different from above. *(Enter in exposure incident screen if this was location where exposure occurred.)*

Address Line 1 _____

Address Line 2 _____

City _____

State ___ ___   Zip ___ ___ ___ ___ ___

County _____   FIPS ___ ___ ___

Sex	☐ Male ☐ Female ☐ Other ☐ Unknown	
Race	☐ 1 Am Indian ☐ 2 Asian/Pacific Is. ☐ 3 Black ☐ 5 White ☐ 6 Mixed	
	☐ 8 Other ☐ 9 Unknown	
Hispanic	☐ Yes ☐ No ☐ Unknown	Speaks English? ☐ Yes ☐ No

Rev. 7/1/2004

APPENDIX C ■ SECTION 3
DATA COLLECTION FORM

Event Information Screen—Event Summary—Application Information
(Complete after interview.)

Event ID __ __ __ __ __ __ __ __ __   Event date ___/___/___   County_____

Event descriptor _____
(Maximum 30-character name for event.)

Licensed applicator ☐
(Use code to indicate level of applicator supervision.)
1=Licensed applicator
2=Licensed trainee, direct supervision
3=Unlicensed, intermittent supervision
4=Unlicensed
8=Not applicable
9=Unknown

Label/use ☐
(Use code to indicate if evidence indicates that the label directions were not followed.)
1=Yes, there was evidence that label directions were not followed.
2=No, no evidence of label directions not being followed.
8=Not applicable
9=Unknown

Application site ☐☐ (Enter code)

Code	Site	Code	Site
01	Farm (excluding, nursery, livestock, forest)	32	Farm product warehousing and storage
02	Nursery	33	Food manufacturing
03	Forest	39	Other manufacturing facility/industrial facility/warehouse facility
04	Livestock and other animal specialty production facility	40	Office/business (non-retail, non-industrial)
05	Greenhouse	41	Retail establishment
09	Other nonproduction agricultural facility	42	Service establishment
10	Single family home	43	Pet care services and veterinary facilities
11	Mobile home	50	Road/rail
12	Multi-unit housing (apartments, multiplexes)	51	Road, rail, or utility right-of-way
13	Labor housing	52	Park
20	Residential institution (dorms, shelters)	54	Private vehicle
21	School	55	Public transportation vehicle
22	Day care facility (incl. in private residence)	59	Other
23	Prison	60	Emergency response vehicle
24	Hospital	70	More than one site
29	Other institution	98	Not applicable
30	Pesticide manufacturing/formulation facility	99	Unknown

Application equipment ☐☐ (Enter code)

Code	Equipment	Code	Equipment
01	Aerial application equipment	10	Trigger pump/compressed air
02	Chemigation	11	Ground sprayer
03	Pressurized can	12	Manual placement
04	Aerosol generator/fogger	13	Dip tank or tray
05	Soil injector	14	More than one type of equipment
06	High-pressure fumigator	15	Other
07	Hand-held granular/dust application	98	Not applicable
08	Spray line, hand-held	99	Unknown
09	Sprayer, backpack		

Rev. 7/1/2004

APPENDIX C ∎ SECTION 3
DATA COLLECTION FORM

Application Target (mark one)

- ☐ (060) Aquatic (pond, stream, lake, irrigation canal)
- ☐ (800) Bait for rodent, bird or predator
- ☐ (200) Beverage crops
- ☐ (041) Building structure (including crack and crevice treatment.)
- ☐ (042) Building surface
- ☐ (043) Building space treatment
- ☐ (530) Cereal grain crops (e.g., barley, corn, wheat, rice)
- ☐ (650) Crops that cross categories 90–600 (general farming)
- ☐ (801) **Community-wide application target**
- ☐ (501) Fiber crops (e.g, cotton)
- ☐ (300) Flavoring and spice crops
- ☐ (510) Forage, fodder hay, silage grasses, silage legumes, and related crops
- ☐ (020) Forest trees and forest lands
- ☐ (100) Fruit crops
  - ☐ (110) Tree fruits
    - ☐ (111) Citrus fruits (e.g., grapefruit, kumquat, lemon, oranges)
    - ☐ (113) Pome fruits (e.g., apples, pears, quince, Japanese plum)
    - ☐ (101) Small fruits (e.g., berries, currants, grapes)
    - ☐ (114) Stone fruits (e.g., apricots, cherries, dates, mangoes, olives)
  - ☐ (120) Subtropical/ other fruits (e.g., avocado, banana, coconut)
  - ☐ (112) Tree nuts (e.g., almonds, hazelnuts, pecans)
- ☐ (500) Grains, grasses, and fiber crops
- ☐ (700) Human
- ☐ (701) Human—skin/hair
  - ☐ (702) Human—clothing
  - ☐ (703) Human—skin/hair and clothing

- ☐ (010) Landscape/ornamental
- ☐ (550) Miscellaneous field crops
- ☐ (600) Oil crops
- ☐ (850) Other (e.g., mixed crop and noncrop, mammal feeding and nesting areas, boats and docks)
- ☐ (601) Seed treatment (application to seeds)
- ☐ (070) Soil
- ☐ (540) Sugar crops (e.g., sugar cane, sorghum)
- ☐ (050) Undesired plant (the plant is the target pest)
- ☐ (400) Vegetable crops
  - ☐ (410) Curcubit vegetables (e.g., cucumbers)
  - ☐ (420) Fruiting vegetables (e.g., cantaloupe, melon, squash)
  - ☐ (430) Leafy vegetables (e.g., cabbage, celery, endive, lettuce)
  - ☐ (460) Other vegetables (e.g., broccoli, cauliflower, eggplant)
  - ☐ (440) Root and tuber vegetables (e.g., beets, carrots, onions)
  - ☐ (450) Seed and pod vegetables (e.g., beans, chick-peas, lentils, peanuts, peas, soybeans, sweet corn)
- ☐ (032) Veterinary/domestic animal
- ☐ (031) Veterinary/livestock
- ☐ (080) Wood product (e.g., utility poles decking, fencing, boardwalk, railroad ties, bulwarks, pilings)
- ☐ (998) Not applicable, application not involved
- ☐ (999) Unknown

Rev. 7/1/2004

APPENDIX C ■ SECTION 3
DATA COLLECTION FORM

Application purpose	☐ 1 Agricultural pest eradication	☐ 8 NA
	☐ 2 Public health pest control or eradication	☐ 9 Unknown

Specific pest target of community-wide application ☐☐☐ (Enter code)

001	Mosquito (no disease specified)	103	Japanese beetle
002	West Nile virus	104	Imported fire ant (red or black)
003	St. Louis encephalitis	105	Asian longhorn beetle
004	Eastern equine encephalitis	106	Emerald ash borer
005	Western equine encephalitis	107	Grain fungal diseases (e.g. black stem rust)
006	La Crosse encephalitis	108	Grasshopper/Mormon cricket
007	Dengue fever	888	Default if State chooses not to code this variable.
100	Boll weevil	996	Multiple pests
101	Gypsy moth (Asian or European)	998	Not applicable (APPTARGT not=801)
102	Fruit fly (Mediterranean, Mexican, Oriental, olive, etc.)	999	Unknown

Event Information Screen—Location

Location where the exposure **event** occurred. This address is the site of the pesticide application, spill, or release (that is, field, orchard, business, institution, residence, or roadway). For locations without specific addresses, include closest crossroad and distances. (This may differ from a person's location at the time of exposure. For example, the exposed person might be located at a school, and the actual event is a fire at a nearby pesticide storage facility. The event location is the pesticide storage facility.)

Address _____
Address _____
City _____
State ___ ___   ZIP ___ ___ ___ ___ ___
Latitude   Longitude
County name _____   FIPS ___ ___ ___

Event Information Screen—Comments—Application/Release Event Narrative

Do not describe exposure. Describe the use/event that involved the release of pesticide. Include details of spill, application, accidental release, etc., that will help clarify how exposures came about.

_____
_____
_____
_____

APPENDIX C ■ SECTION 3
DATA COLLECTION FORM

Event Information Screen—Pesticide Products

If the EPA registration number is known, complete the EPA registration number and product name below then skip to **Chemical Agent Comments** on page 8. (Active ingredient code [PC Code, percentage, form, chemical class, and functional class are auto-entered in SPIDER; record only if using lookup file for entry into nonautomated system.) If the EPA registration number is unknown, but the identity of active ingredient(s) is known, enter the most detailed product name available below and all other available information regarding the ingredients on the table labeled **Active Ingredient** on page 7. Information about products where only form and functional class are known, or carriers and inerts can be entered in the section labeled **Other Source** on page 8.

Record all information available including manufacturer and any modifiers on label (e.g., spray, dust, 4E).

EPA registration number/distributor number	Name	Form*	Poisoning attribution†
a. _____ - _____ / _____	_____	____	☐
b. _____ - _____ / _____	_____	____	☐
c. _____ - _____ / _____	_____	____	☐
d. _____ - _____ / _____	_____	____	☐
e. _____ - _____ / _____	_____	____	☐

*See form codes on next page.
†*Check box if product is thought to have contributed to illness. Complete this column at time of case closure.*

Rev. 8/10/04

APPENDIX C ■ SECTION 3
DATA COLLECTION FORM

Event Information Screen, Pesticide Product—Active Ingredients and Other Sources
Active Ingredient

Active ingredient code	Name	Per-centage	Form*	Chemical Class*	Functional Class*	Poisoning Attribution[†]
a.						☐
b.						☐
c.						☐
d.						☐

*Indicate the product form, chemical, and functional class from the tables that follow.
[†]Check box if product is thought to have contributed to illness. Complete this column at time of case closure.

NIOSH form codes			
01	Dust/powder (not pressurized)	10	Flowable concentrate
02	Granular/flake	11	Pressurized liquid/spray/fogger
03	Pellet/tablet/cake/briquette	12	Ready-to-use liquid/solution
04	Wettable Powder/Dust	13	Other liquid formulation
05	Impregnated material (ant/plant stakes, animal collars, water filters)	14	Pressurized gas/fumigant
		15	Paint/liquid coating
06	Other dry formulation	16	Other
07	Microencapsulated	17	Soluble powder
08	Emulsifiable concentrate	18	Liquid concentrate
09	Soluble concentrate	99	Unknown

Chemical class codes	Functional class codes
01 Organochlorine compound	01 Insecticide (excluding solely IGR and fumigants)
02 Organophosphorous compound	02 Insect growth regulator (IGR)
03 N-methyl carbamates	03 Herbicide/algicide
04 Pyrethrin	04 Fungicide
05 Pyrethroid	05 Fumigant
06 Dipyridyl compound	06 Rodenticide
07 Chlorophenoxy compound	07 Disinfectant/broad spectrum for water sanitation
08 Triazines	08 Insect repellent
09 Carbamates (non-AChE inhibitors)	09 Antifouling agent (marine paints)
10 Organo-metallic compound	10 Insecticide and herbicide (01 & 03)
11 Inorganic compounds	11 Insecticide and fungicide (01 & 04)
12 Coumarins	12 Insecticide and herbicide and fungicide (01, 03, & 04)
13 Indandiones	13 Insecticide and other (01 & 96)
14 Convulsants	14 Herbicide and fungicide (03 & 04)
15 Microbial	96 Other (includes biological controls, plant growth regulators, antibiotics, etc.)
16 Dithiocarbamates	
95 Unidentified cholinesterase inhibitor	97 Multiple (product is classified as multiple classes which do not fit in any of the codes specified in codes 10–14)
97 Multiple (PC Code indicates a code for a combination of active ingredients that cross chemical classes)	99 Unknown
99 Unknown	

Rev. 8/10/04

APPENDIX C ■ SECTION 3
DATA COLLECTION FORM

Other Source If neither product nor chemical ingredient known.

Enter a description of the other pesticide, e.g., "Some kind of spray from a highway truck"; "unspecified Black Flag wasp spray"; "unlabeled spray can." This area can also be used to record information about carriers and inerts at the State level.

Other ID	Description of other source	Chemical class*	Form*	Functional class*	Poisoning attribution[†]
_____	_____	_____	_____	_____	☐
_____	_____	_____	_____	_____	☐
_____	_____	_____	_____	_____	☐

* *Indicate the product form, chemical, and functional class from the tables above.*
[†]*Check box if product is thought to have contributed to illness. Complete this column at time of case closure.*

Event Information—Comments—Chemical Agent Comments
(Note additional information about pesticide products and adjuvants.)

_____
_____
_____
_____
_____
_____
_____
_____
_____
_____
_____
_____
_____
_____
_____
_____
_____
_____
_____

Rev. 8/10/04

APPENDIX C ■ SECTION 3
DATA COLLECTION FORM

Exposure Information Screen—Incident Information
Incident Report Information

Exposure ID	Report date __ __ / __ __ / __ __ __ __
Case ID __ __ __ __ __ __	Event ID __ __ __ __ __ __

Report source 1 __ __ (__)   **Report source 2** __ __ (__)   **Report source 3** __ __ (__)

Use codes for sources below. Note that an additional character can be added for State-specific codes under each category (e.g., listing specific poison control centers in the State by using codes 02A-02Z or 020-029).

Source codes

Code	Description
01	Physician report
02	Poison control center
03	Other health care provider report (including ER or hospital report)
04	Laboratory report
05	Death certificate or medical examiner's report
06	Report or referral from governmental agency
07	Obituary/news report
08	Ascertainment through Worker's Compensation
09	Self-report
10	Co-worker report
11	Friend or relative report
12	Identified during site visit
13	Worker representative (e.g., union, lawyer/legal services/other advocate)
14	Medical record review (clinic or hospital record review performed by surveillance staff)
97	State Department of Health
98	Other (not captured in any code category listed)
99	Unknown

Site of exposure ☐ ☐ *(Enter code)*

01	Farm (excluding, nursery, livestock, forest)	32	Farm product warehousing and storage
02	Nursery	33	Food manufacturing
03	Forest	39	Other manufacturing facility/industrial facility/warehouse facility
04	Livestock and other animal specialty production facility	40	Office/business (nonretail, nonindustrial)
05	Greenhouse	41	Retail establishment
09	Other nonproduction agricultural facility	42	Service establishment
10	Single family home	43	Pet care services and veterinary facilities
11	Mobile home	50	Road/rail
12	Multiunit housing (apartments, multiplexes)	51	Road, rail, or utility right-of-way
13	Labor housing	52	Park
20	Residential institution (dorms, shelters)	54	Private vehicle
21	School	55	Public transportation vehicle
22	Day care facility (including in private residence)	59	Other
23	Prison	60	Emergency response vehicle
24	Hospital	70	More than one site
29	Other institution	98	Not applicable
30	Pesticide manufacturing/formulation facility	99	Unknown

Rev. 8/10/04

APPENDIX C ■ SECTION 3
DATA COLLECTION FORM

Activity of case at time of exposure ☐☐ *(Enter code)*
01 Applying pesticide
02 Mixing/loading pesticide
03 Transport or disposal of pesticide
04 Repair or maintenance of pesticide application equipment
05 Any combination of activities 01–04
06 Involved in manufacture or formulation of pesticide
07 Emergency response
08 Routine work activity not involved with pesticide application (includes exposure to field residue)
09 Routine indoor living activities not involved with pesticide application
10 Routine outdoor living activities not involved with pesticide application
98 Not applicable
99 Unknown

Others exposed

Were other persons possibly exposed? ☐ Yes ☐ No ☐ Unknown

If Yes, How many? _____

Did any seek medical care? ☐ Yes ☐ No ☐ Unknown

Use a separate sheet of paper to record names and contact information if appropriate.

Exposure address *(That is, subject's location at time of exposure. This may be the same as the case address or the event address.)*

Address 1 _____
Address 2 _____
City _____
State __ __   ZIP __ __ __ __ __
FIPS __ __ __   County __ __ __
County name _____

Note: For cases reported multiple times, you can use the shortcut buttons in SPIDER to either copy the existing address from the case table, or if this is a new address, move this address to the case table.

Rev. 8/10/04

APPENDIX C ■ SECTION 3
DATA COLLECTION FORM

Initial treatment /HCP information

First care	Where was medical care first sought? ☐ *(Enter code)*
	1=Physician's office 5=No medical care sought
	2=Emergency room 6=Other
	3=Hospital admissions 7=Employee health center
	4=Advice of poison control center 9=Unknown

HCP name   Last _____

               First _____

Type	Who provided care? ☐ *(Enter code)*
	1=Family physician 4=Consulting specialist
	2=Employer's physician 5=Outpatient clinic
	3=Worker's Comp physician 9=Unknown

	Chart location	Work location
Address 1	_____	_____
Address 2	_____	_____
City	_____	_____
State	__ __	__ __
ZIP	__ __ __ __ __	__ __ __ __ __
Phone	(__ __ __) __ __ __ - __ __ __ __	(__ __ __) __ __ __ - __ __ __ __

Exposure Information Screen—Medical Information

☐ **Diagnosis made by HCP** *(Checked=Yes)*

     Diagnosis: _____

     ICD9: __ __ __ . __ __

     Summary: _____

☐ **Hospitalized** *(Checked=Yes)*

     Admit date: __ __ / __ __ / __ __   **Discharge date**: __ __ / __ __ / __ __

     Length of stay: __ __ __ (days)   Coding   997 if > 996 days

                                                                                               998=NA, not hospitalized

                                                                                              999=Unknown

Rev. 8/10/04

APPENDIX C ■ SECTION 3
DATA COLLECTION FORM

Facility where hospitalized

Facility name _____

Facility address _____

Treating physician _____

Condition present at time of exposure? *(Circle one for each condition.)* **Coding for conditions**

Pregnant	1	2	3	4	5	9	1=Doctor reported
Allergies	1	2	3	4	5	9	2=Exposed person reported
Asthma	1	2	3	4	5	9	3=Both doctor and person reported
							4=Condition was absent
Acquired chemical intolerance	1	2	3	4	5	9	5=Not reported
							9=Unknown

Other *(Enter condition and code.)* _____

Outcome ☐	1= Fatal, pesticide-related 8=Not applicable (not fatal) 2= Fatal, not pesticide-related 9=Unknown 3=Fatal, relation unknown
Lost time ☐	1=Yes, one or more days lost from work 2=No, no time lost 3=Unemployed, lost 1 or more days from school or regular activities 9=Unknown
Total time lost	___ ___ ___ . ___ ___ days
Followup needed?	☐ Check if yes. When? ___ ___ / ___ ___ / ___ ___

Non-Cholinesterase Chemical-Specific Biological Test for Pesticides or Metabolites

Other biological tests?	☐ Yes	☐ No	☐ Unknown	
	Test 1		**Test 2**	
Test type				
Draw date	___ ___ / ___ ___ / ___ ___		___ ___ / ___ ___ / ___ ___	
Numeric result				
Analysis result	☐ Abnormal ☐ Not applicable	☐ Normal ☐ Unknown	☐ Abnormal ☐ Not applicable	☐ Normal ☐ Unknown

Notes _____

Rev. 8/10/04

APPENDIX C ■ SECTION 3
DATA COLLECTION FORM

Exposure Information Screen—Cholinesterase Results
Coding guidance for completing table of results
Option 1 Detailed version.
Option 2 Required minimum. *Enter single test response for* **Test type** *and* **Result type**, *the only required fields.*

PFI	Lab code from lab pick list or enter lab name						
Test type codes	1=RBC 2=Plasma 5=Either RBC or Plasma		3=Both RBC and Plasma 8=Not applicable			4=Not done 9=Unknown	
Result type codes	1=Abnormal compared to lab 3=Normal compared to lab 7=Bad specimen		2=Abnormal compared to baseline 4=Normal compared to baseline 8=Not applicable			9=Unknown	
PFI	Lab Name	Test Type	Test Date	Numeric Result	Result Type	Lab Low	Lab High
			__/__/__				
			__/__/__				
			__/__/__				
			__/__/__				

Exposure Information Screen—Narrative
Brief exposure description (120 characters)

_____
_____

Exposure narrative *Include details of person's activity or situation that resulted in the exposure. Include notes on medical, exposure, and follow-up activity information. Continue on a separate sheet if necessary.*

_____
_____
_____
_____
_____
_____
_____
_____
_____

Rev. 8/10/04

APPENDIX C ■ SECTION 3
DATA COLLECTION FORM

Exposure Information Screen—Nature of Exposure

Type of exposure			Route of exposure		
[ ] Drift	[ ] Indoor Air	[ ] Contact	[ ] Dermal	[ ] Ingestion	[ ] Ocular
[ ] Spray	[ ] Surface	[ ] Unknown	[ ] Inhalation	[ ] Injection	[ ] Unknown
[ ] Other					

Intentional? ___   Coding: 1=Yes, suspected intentional   2=No, unintentional   9=Unknown

Date first exposed* __ __/__ __/__ __   Time __ __ __ __ (24 Hour Clock)

Date symptom onset* __ __/__ __/__ __   Time __ __ __ __ (24 Hour Clock)

Other date   __ __/__ __/__ __   Other Date Description_____

* At least one of the following dates must be entered: first exposure, symptom onset, or laboratory test (page 12).

Exposure Information Screen—Narrative

Date comments. *Indicate any notes on date of exposure onset, report, or lab test pertinent to understanding case chronology.*

_____
_____
_____
_____

Exposure Information Screen—Occupational Information

Work related? ☐   1=Yes   2=Possibly   3=No   4=Unknown   5=Not Applicable
If answer is 3 or 5, skip to PPE Use below.

Job title 60 characters	COC title
_____	_____
_____	_____
Occupation narrative 125 Characters	**CIC title**
_____	_____
_____	_____
Industry at time of exposure 100 Characters	
_____	

Rev. 11/22/04

APPENDIX C ■ SECTION 3
DATA COLLECTION FORM

Employer Information—Employer Screen

Employer ID _____   Name _____

Address Line 1 _____

Address Line 2 _____

City _____

State ___ ___   ZIP ___ ___ ___ ___ ___

Phone (___ ___ ___) ___ ___ ___ - ___ ___ ___ ___

Contact _____

NAICS Code ___ ___ ___ ___ ___ ___

Exposure Information Screen—Occupational Information, PPE Use

PPE Use   *(Complete this section after interview using codes below.)*

1=Used (all or some of PPE required)
2=Used (not required)
3=Used (unknown requirements)
4=Not used (some PPE required)
5=Not used (unknown requirements)
6=Not used (not required)
8=Not applicable
9=Unknown

Specific PPE Used ☐ ☐

Codes: 1=Yes, used   2=No, not used   8=Not applicable   9=Unknown
Check one box for each form of PPE.

	1	2	8	9		1	2	8	9
Supplied Air	☐	☐	☐	☐	Natural Gloves	☐	☐	☐	☐
Respirator	☐	☐	☐	☐	Synthetic Gloves	☐	☐	☐	☐
Dust Mask	☐	☐	☐	☐	Goggles	☐	☐	☐	☐
Boots	☐	☐	☐	☐	Engineering	☐	☐	☐	☐
Clothing	☐	☐	☐	☐					

Worker Protection Standard

If agricultural worker or pesticide handler (farm, nursery, or forestry), for each question indicate if response is (Y)Yes (N) No (U) Unknown/Not asked by circling the letter for the response.

a. Did this incident involve entering a treated area (including field or greenhouse)?   Y   N   U

b. If yes, did your employer/crew leader tell you how soon you could go into the area after it was treated?   Y   N   U

Rev. 11/22/04

APPENDIX C ■ SECTION 3
DATA COLLECTION FORM

Exposure Information Screen—Signs and Symptoms

*Fill in **Doctor reported** column based on medical records or HCP interview. Final code column should be completed prior to data entry. Codes are listed following the table on page 18. Check all signs or symptoms described, or stated as absent (items in italics should be taken from HCP interview or medical record only).*

System	Sign/symptom	Patient reported	Doctor reported	Absent	Final code
General	*Acidosis*				
	Alkalosis				
	Fatigue/malaise				
	Fever				
	Increased anion gap				
	Other_____				
Cardiovascular	*Bradycardia*				
	Cardiac arrest				
	Chest pain				
	Conduction disturbance				
	Hypertension				
	Hypotension				
	Palpitations				
	Tachycardia				
	Other_____				
Renal	Frequent urination				
	Oliguria/anuria				
	Blood in urine				
	Proteinuria				
	Other_____				
Neurological	Altered taste				
	Anxiety/hyperactivity/irritability				
	Ataxia /trouble walking				
	Blurred vision				
	Coma				
	Confusion				
	Diaphoresis (profuse sweating)				
	Dizziness				
	Fainting				
	Headache				
	Memory loss				
	Muscle pain				
	Muscle rigidity				
	Muscle twitching/fasciculations				

Rev. 8/10/04

APPENDIX C ■ SECTION 3
DATA COLLECTION FORM

System	Sign/symptom	Patient reported	Doctor reported	Absent	Final code
Neurological (continued)	Muscle weakness				
	Paralysis				
	Paresthesias/tingling or numbness				
	Peripheral neuropathy				
	Salivation				
	Seizure				
	Slurred speech				
	Other_____				
Gastrointestinal	Anorexia (loss of appetite)				
	Constipation				
	Diarrhea				
	GI bleeding (blood in stool or vomit)				
	Nausea				
	Pain				
	Vomiting				
	Other_____				
Eye	*Burns*				
	Conjunctivitis (diagnosis)				
	Corneal abrasion				
	Miosis				
	Mydriasis				
	Pain/irritation/inflammation				
	Tearing/*lacrimation*				
	Other_____				
Dermal	Blisters/*bullae*				
	Burns				
	Edema/swelling				
	Hives				
	Pain				

Rev. 8/10/04

APPENDIX C ■ SECTION 3
DATA COLLECTION FORM

System	Sign/symptom	Patient reported	Doctor reported	Absent	Final code
Dermal (continued)	Pattern* of rash or lesions				
	Pruritis (itching)				
	Rash				
	Redness				
	Other_____				
Respiratory	*Asthma (diagnosis of)*				
	Cough				
	Cyanosis				
	Depression				
	Difficulty breathing/ shortness of breath				
	LR irritation				
	Pleural pain (pain on deep breathing)				
	Pulmonary edema				
	Tachypnea				
	UR irritation				
	Wheezing				
	Other_____				

*Coding for pattern of dermal lesions
1=Corresponds well with physical pattern of exposure
2=Discrete patches of lesions do not correspond with the pattern of exposure
3=Generalized distribution of lesions on the body
4=Absent
9=Unknown

Final Codes for All Fields 1=Doctor reported   2=Exposed person reported   3=Both Dr. and person reported
9=Unknown

Exposure Information Screen—Narrative

Health comments _____

_____

_____

_____

_____

_____

Rev. 8/10/04

APPENDIX C ■ SECTION 3
DATA COLLECTION FORM

Exposure Information Screen—Medical Staff

Medical ID	_ _ _ _ _ _ _ _ _ _ _ _ _ _ _ _ _ _ _ _ _ _ _ _

Enter one Medical ID for each medical person involved in the case. Use F2 in SPIDER to select from pick list. See HCP Information on Page 11 and enter on Medical Staff screen. If not on pick list, complete the full medical staff information.

Case Tracking and Closure

Event Information Screen—Event Summary

Violations FIFRA	1 2 3 4 8 9	Were citations for violations of regulations by these agencies issued?
Violations OSHA	1 2 3 4 8 9	1=Violation cited 8=Decision not to refer 2=No violation cited 9=Unknown
Violations Other	1 2 3 4 8 9	3=Citation pending 4=Case refused referral to agency

Referrals	Check if referral made	Date of referral (MM/DD/YY)	Notes
IH Staff			
Ag and Mkt Program			
State Environment			
EPA			
Ag Program			
Other State Pgm.			
Local Health Unit			

Event Information Screen—Comments

Other violation descriptions _____

_____

_____

Violation comments _____

_____

_____

Rev. 8/10/04

APPENDIX C ■ SECTION 3
DATA COLLECTION FORM

Exposure Information Screen—Exposure and Classification

Severity ☐
        1 = Fatal   2 = High   3 = Moderate   4 = Low   8 =Evaluated, not applicable

A. Documentation of exposure ☐ ☐
(Put a number in the first box and letter in the second box if appropriate.)

1 - Confirmed by   a-envir/bio testing   b-professional observation   c-biological evidence
                    d-eye/derm signs   e. 2+ findings by medical staff

2 - Reported by   a-case   b-witness   c-application records
                 d-nonprofessional observation   e-other

3 - Strong evidence of no exposure

4 - Insufficient data

B. Documentation of health effect ☐
    1 - 2+ Findings by medical staff
    2 - 2+abnormal symptoms
    3 - No post exposure findings
    4 - Insufficient Data

C. Evaluation of causal relationship ☐ ☐

(Put a number in first box and letter in second box if first box is 1.)
    1 - Fits known toxicology
        a-characteristic (Appendix 2 of case classification) and temporal relationship is plausible
        b-consistent with literature and known toxicology

    2 - Inconsistent with known toxicology

    3 - Definitely ruled out (evidence of non-pesticide causal agent)

    4 - Insufficient toxicologic information available

NIOSH Classification ☐   Alternate Classification ☐

Classification categories   1=Definite   5=Unlikely
                                2=Probable   6=Insufficient Information
                                3=Possible   7=Exposed/Asymptomatic
                                4=Suspicious   8=Unrelated

Exposure Information Screen—Poisoning Attribution

Return to pages 6 - 8 to determine if illness is attributable to products, active ingredients, or substances listed there.

Rev. 8/10/04

APPENDIX C ■ SECTION 4

C.4 FIELD INVESTIGATION CONTACT FORM AND HEALTH SAFETY CHECKLIST FOR FIELD PERSONNEL

The following checklist is designed to remind field staff about equipment and safety precautions they need to take when embarking on a field investigation or any form of on-site follow-up. It also ensures that supervisory staff have appropriate contact information for field staff. In addition, PPSPs may decide to use this form when staff are accompanying partner enforcement agency staff. The checklist is to be completed by PPSP personnel BEFORE going out into the field. For team visits, only the team leader is required to complete the form. Team leaders are responsible for ensuring that all team members meet the required checklist items stated below. As with all of the sample forms, this form should be modified to meet the specific needs of the PPSP.

APPENDIX C ■ SECTION 4

Field Investigation Contact Form and Health Safety Checklist for Field Personnel
(Adapted from California Occupational Health Branch, DHS)

This checklist is to be completed by PPSP personnel BEFORE going out into the field. For team visits, *only* the team leader is required to complete the form. Team leaders are responsible for ensuring that all team members meet the required checklist items stated below.

Directions
1. Fill out **Part I**.
2. Have it reviewed and signed by your supervisor **before** the site visit.
3. After the visit, fill out **Part II** and turn in completed form to supervisor.
4. **Supervisors:** Forward a copy of this form to_____.

Part I: Pre-Site Visit Checklist

A. Names of PPSP field team members
_____
_____
_____

    Date(s) of proposed site visit_____

B. Emergency Contact Information

    Employer name _____

    Type of business _____

    Address _____

    City _____

    Contact person: Name_____ Phone (____)_____

How can you be reached?

Phone number where you are staying (e.g., hotel friend's house, etc.) (____)_____

Hotel name *(if applicable)* _____

Cellular Phone number *(if applicable)* (____) _____

If more than one location, write down team members' locations *(use back of page if necessary)*.

Phone (_____) _____

Hotel name *(if applicable):* _____

Cellular phone no. *(if applicable)* (____) _____

APPENDIX C ■ SECTION 4

Training (hazards, guidelines, regulations)

(1) Will you be potentially exposed to the following hazards? *(Check all that apply.)*

☐ TB: ➡ PPD of staff performed in past year? ☐ yes ☐ no

☐ Other chemical(s) _____
_____

☐ Noise _____

☐ Bloodborne pathogens _____
➡ HBV vaccination completed? ☐ yes ☐ no

☐ Safety (falling objects, electricity, etc.): _____

☐ Violence _____

☐ Other: _____

(2) If you checked any of the boxes above, have you had training on how to protect yourself from these hazards? ☐ yes ☐ no
➡ *If you answered NO, please talk to your supervisor about how you will obtain appropriate training prior to the site visit.*

(3) Have you reviewed applicable regulations and guidelines on likely exposures?
☐ yes ☐ no
➡ *If you answered NO, please review any applicable regulations and guidelines that are available. If none exist, you should discuss alternative information sources with your supervisor.*

Personal Protective Equipment (PPE)

(1) Are respirators required or recommended on this site visit?
☐ yes ☐ no ☐ not sure
➡ *If NOT SURE, please discuss this with your supervisor.*

If you are using a respirator for this site visit, answer a through c.
(a) Have team members had

Respirator medical clearance within the last year?	☐ yes	☐ no
Respirator training within the last year?	☐ yes	☐ no
Respirator fit-testing within the last year?	☐ yes	☐ no

➡ *If NO, please discuss this with your supervisor.*

163

(b) Do you have extra cartridges/filters? ☐ yes ☐ no

If you are going to wear a PAPR, have you charged the battery packs?
☐ yes ☐ no

(2) Are other PPE required? ☐ yes ☐ no, If YES, check all that apply:

☐ head protection (hard hat)   ☐ foot protection (safety shoes or chemical resistant boots)
☐ hearing protection   ☐ eye protection (goggles, faceshield)   ☐ hand protection
☐ other

Communications Equipment

(1) If you are traveling to remote areas, do you need a cellular phone? ☐ yes ☐ no
(2) Do you have a list of emergency contact numbers (e.g., section chief, etc.) to bring with you? ☐ yes ☐ no

Employee signature: _____ Date: _____

Supervisor signature: _____ Date: _____

Part II: Post-Site Visit Review

Were there any items missing or not foreseen prior to visit that should be considered during future visits to similar sites? ☐ yes ☐ no

➡ *If YES, please explain below.*

C.5 Instructions for National Transportation Safety Board (NTSB) Search to Obtain Reports of Airplane Accidents Involving Aerial Pesticide Applicators

NTSB maintains data on aviation accidents. The Aviation Synopses can be searched for information about aviation accidents involving aerial agricultural applications. These applications are covered by 14 CFR Part 137. The synopses may be reviewed by searching monthly lists of accidents or performing a query. Make sure your browser is set to accept cookies.

Go to the NTSB website http://www.ntsb.gov/NTSB/query.asp.

The first item on the list is a *Database Query*. First, look at the general instructions for searching, then select the *Database Query* form. Complete the various fields. This includes entering the date range you are interested in, selecting your State, severity, and a specific category of aircraft (if you are searching for a particular accident or type of accident). You can leave the *Operation* category as *All*. In the area labeled *Enter your word string below*, type *137*. Choose *Sort by Date Ascending* unless you want to sort on an option other than date. Click on *Submit Query*. This search strategy should catch all of the agricultural accidents (since applications are regulated by 14 CFR Part 137).

C.6 Sample Templates for Tables Presenting Surveillance Data

1. Work-Relatedness by Case Classification Status

	Definite	Probable	Possible	Suspicious	Total
WORKREL					
Yes					
Possibly					
No					
Unknown					
Total					

2. Cases by Pesticide Type and Case Classification.

Each case should be included only once in this table (excluding the rows and columns of totals). (Note that this tabular presentation can be done separately for occupational and nonoccupational cases or can include occupational status by splitting the case classification columns into occupational and nonoccupational.)

	Definite	Probable	Possible	Suspicious	Total
Pesticide type					
Insecticides-total					
Cholinesterase inhibitors					
Pyrethrin/pyrethroid					
Other insecticides					
Insect growth regulators					
Herbicide/algicide					
Fungicide					
Fumigant					
Rodenticide					
Disinfectant					
Insect repellent					
Other					
Multiple					
Unknown					

3. Occupational Pesticide Injury and Illness Cases by Occupation and Case Classification

	Definite	Probable	Possible	Suspicious	Total
Occupation					
Agriculture—provide total for COC codes 473–499					
List specific agriculture occupations					
Nonagriculture—provide total					
List specific nonagriculture occupations					
TOTAL					

4. Source of case report by gender and case classification

	Definite			Probable			Possible			Suspicious			Total		
Sex	M	F	U	M	F	U	M	F	U	M	F	U	M	F	U
Source of Report															
Physician															
Poison control center															
Other health care professional															
Death certificate															
Government agency															
Media report															
Workers Compensation															
Self-report (incl. relative or co-worker)															
Employer															
Other															
Unknown															
Total															

5. Age of cases by gender and case classification (produce this table for occupational and nonoccupational exposure status)

	Definite			Probable			Possible			Suspicious			Total		
Sex	M	F	U	M	F	U	M	F	U	M	F	U	M	F	U
Age															
<10															
11–14															
15–17															
18–29															
30–39															
40–59															
60–79															
80+															
Unknown															
Total															

C.7 Sample Letters for PPSP Case Follow-up

The following sample letters are provided as templates that PPSPs can modify to meet their specific needs and legal requirements.

- Thank-you letter to an HCP who reports a case
- A request for cooperation to an HCP who failed to report a case
- A request to an HCP for medical records
- A letter to an employer regarding an upcoming site inspection

Thank-You Letter to HCP Who Reported Case

[Agency Letterhead]

DATE

INSIDE ADDRESS

Re: [**case number**]

Dear [**insert HCP name**]:

Thank you for the information you recently provided regarding the illness and possible pesticide exposure of [**insert patient name**]. Your report helps us to identify pesticide products and practices that may affect public health, as well as provide exposure prevention information to affected individuals.

If you would like further information about this case, the [**insert agency name**] [**insert surveillance program name**], or other State agency resources, please call me at [**insert phone number**]. [**Option**—include a copy of the *EPA Recognition and Management of Pesticide Poisonings*]

Sincerely,

Request for Cooperation Letter to HCP Who Failed to Report Case

[Agency Letterhead]

DATE

INSIDE ADDRESS

Re: [**patient**]   Case number:

Dear [**insert HCP name**]:

The [**insert agency name**] is currently investigating a reported pesticide-related illness of your above-named patient. (We have been in touch with a member of your staff regarding the observable signs, diagnosis, and treatment of the individual.) The [**surveillance program name**] routinely identifies and investigates illnesses and injuries associated with pesticide exposure.

Suspected pesticide poisoning is a reportable condition in the [**insert State name**]. Health care providers are required by [**insert rule reference**] to report acute or subacute conditions that are caused by, or suspected of being caused by, pesticide exposure. All medical details and the person's identity are kept confidential. Resources and referrals are available to the reporting provider and the patient, including exposure prevention information.

Your report helps us identify pesticide products or practices that may affect public health. Your cooperation in reporting any future pesticide-related illnesses is appreciated. If you would like further information about the program, please contact us at [**insert phone number**]. [**Option—include a copy of the** *EPA Recognition and Management of Pesticide Poisonings*.]

Sincerely,

MEDICAL RECORDS REQUEST LETTER

[Agency Letterhead]

DATE

INSIDE ADDRESS

Re: [**case number**]

Dear [**insert HCP name**]:

This letter is sent by [**agency name (agency abbreviation)**] to request medical records relevant to illness investigations conducted by the [**agency abbrev.**]. The [**agency**] collects medical records in accordance with State law. Copies of relevant sections of State code are attached for your convenience.

The [**agency**] has received a notice of pesticide-related illness involving the patient listed below and requests copies of any medical information (including chart notes and laboratory test results) that you might have.

[**First name Last name, SSN, DOB; injured on: date; seen on: date**]

Please also check your records for any information you have on other patients seen on the same date, or within several days before or after, who may have a pesticide-related illness or injury associated with the same exposure incident. Please provide copies of any such records to the [**agency**].

Please mail or fax these records to the return address indicated above. If you have any questions, please call me at [**insert phone number**]. Thank you for your cooperation in providing the requested information.

Note that the [**agency name**] is an agency of [**parent authority, e.g., the State of _____**] and is conducting pesticide poisoning surveillance in its capacity as a public health authority as defined by the Health Insurance Portability and Accountability Act (HIPAA), Standards for Privacy of Individually Identifiable Health Information; Final Rule (Privacy Rule) [45 CFR 164.501]. Pursuant to 45 CFR 164.512(b) of the Privacy Rule, covered entities such as your organization may disclose, without individual authorization, protected health information to public health authorities ". . .authorized by law to collect or receive such information for the purpose of preventing or controlling disease, injury, or disability, including, but not limited to, the reporting of disease, injury, vital events such as birth or death, and the conduct of public health surveillance, public health investigations, and public health interventions . . ." The information being requested represents the minimum necessary to carry out the public health purposes of pesticide poisoning surveillance pursuant to 45 CFR 164.514(d) of the Privacy Rule.

Sincerely,

Letter to Employer Regarding an Upcoming Site Inspection

[Agency Letterhead]

DATE

INSIDE ADDRESS

Re: Site inspection scheduled for [**date**]

Dear [**name**]:

Thank you for your cooperation with the [**agency name**] investigation of illness reports among employees at [**company name**]. We are writing to confirm our meeting and to provide you with logistical and other information about our investigation. The meeting will take place at [**insert location**].

Based on phone conversations with [**person's name (date)**], it is our understanding that the following persons will be present at the meeting: [**names and titles**]. In addition to ourselves, [**list any additional names and titles**] working with [**agency and program name**] on pesticide illness tracking will also attend our meeting.

Background: [**agency**] is mandated to investigate the causes of morbidity and mortality from work-induced diseases and develop recommendations for improved control of work-induced diseases [**code reference**] [**Modify sentence as needed to reflect agencies' authority**]. In contrast to [**State**] - OSHA, [**program name**] is not an enforcement agency, and we do not issue citations. [**agency**] initiated this investigation in response to physician reports of illness among [**worker types**] potentially exposed to pesticides as a result of [**work activity or task**]. Between [**dates**], [**agency**] received [**insert number**] incident reports involving a total of [**insert number**] workers. [**Give reason there is concern about the reports.**]

Purpose: The purpose of our on-site field investigation on [**date**] is to gather information and make observations about [**activity**] practices at [**company name**].

Process: The process for the [**date**] site visit will be an opening conference for introductions, a review of the purpose, scope, and methodology of our investigation, and an opportunity for company and worker representatives to ask us questions. We will then conduct an on-site observation of [**process, site, or activity**]. We will walk through the [**facility type**], at which time we will ask [**company name**] representatives to provide a detailed explanation of the pesticide application process including local ventilation conditions at the time of the application. We will ask worker representatives for their knowledge of the process as well. We will take photographs. The site visit will end with a closing conference at which time we will summarize our progress in the investigation and answer any additional questions you may have. We anticipate that the site visit will take about four hours (assuming that we will have received much of the information/documentation about the application from [**company name**] prior to the site visit).

Scope: The scope of [**agency**] investigation is limited to [**describe scope**]. The limited nature of [**agency name's**] investigation does not imply there are, or are not, other safety and health issues at the workplace.

Methodology: [**agency name**] will evaluate and classify [**worker or specific worker job type**] illness reports according to criteria established by the National Institute for Occupational Safety and Health (NIOSH). Enclosed is a copy of these criteria. They are also available at http://www.cdc.gov/niosh/pestsurv/pdfs/pest-casdef2000.pdf. Worker exposure to pesticides will be assessed utilizing data from (1) interviews with employees, employee representatives, and employer representatives; (2) work-site observations and interviews with company and employee representatives regarding the work process, tasks, and exposure control measures; (3) review of medical records, policy and procedures, and other written materials; and (4) a review of the relevant scientific literature. [**Agency**] will attempt to conduct a voluntary interview with all workers with a reported illness at the worker's home by phone. Employer interviews will be conducted at the work-site and by phone as needed.

Exposure control measures will be evaluated according to the presence, use, and efficacy of standard industrial hygiene hierarchy of controls (i.e., engineering, administration, and personal protection). Recommendations to prevent illness will be based on a public health approach; i.e., primary, secondary, and tertiary measures. To the extent possible, it is normal practice for [**agency**] to direct each of our recommendations to the persons or groups that have the authority to implement change.

Our investigation will be conducted independently of regulatory agencies. However, if while at a workplace, we observe a condition that could reasonably be expected to cause death or serious physical harm immediately (that is, an imminent hazard), we are obliged to notify the employer and affected workers of the hazard and to notify [**State name**]-OSHA and/or other appropriate agencies. In practice, circumstances that would require a referral to an enforcement agency have almost never been encountered by [**agency**] investigators. The investigation may not identify all hazards or violations of good practice within the scope of the practices reviewed Allowing [**agency**] to conduct the investigation and/or following recommendations made in the investigation report will not exempt [**company name**] or the worksite from an enforcement inspection or regulatory compliance.

At a minimum, all of [**agency's**] findings and recommendations to prevent illness will be reported in writing in a timely manner to the incident cases, reporting physicians, employee representatives, and [**company name**]. Publications may also be disseminated to other interested parties, such as health and safety professionals, industry-based organizations, government agencies, and labor unions. Our report will not contain any personal-identifying information about individual workers. Although not confidential information, our publications for general distribution do not usually specify the name of the employer.

Specific information we are seeking from [**company name**]: In order to make the best use of your time during our site visit, we are providing you with a list of the questions and information we will request from [**company**]. We will also have additional questions based on what we learn from you. All of the information requested below relates to the time period covered by the scope of this investigation ([**date range**]), except where otherwise indicated.

[list questions]

We have tried to compile a comprehensive list of questions and sources of information that are relevant to this investigation. However, it is likely that we have omitted something. Please do not hesitate to provide any other data that you are aware of that may be useful in understanding the work process of [**process, site, or activity being investigated**]. Also, please note if certain data are not available, as it is important for us to understand where there may be gaps in data.

We appreciate your time and participation in the [**agency**] investigation. It is our goal that the information collected will contribute to our ability to determine the severity and extent of the potential problem and identify possible causes and solutions. Please contact [**name**] by e-mail ([**e-mail address**]) or phone ([**phone number**]) if you have any questions. We look forward to meeting you on [**date**].

Sincerely,

cc:

Enclosures

C.8 Instructions for Obtaining Acute Pesticide-Related Illness and Injury Reports From Poison Control Centers (PCCs)

A. Obtaining the annual number of incident cases
1. Contact your local PCC. Contact information can be obtained from the American Association of Poison Control Centers at http://www.aapcc.org/director2.htm. Some States have more than one PCC.

2. Include in-State residents and those of unknown residence.

3. Determine if the PCC uses the Toxicall® data system.
   a. If YES, go to step A.4.
   b. If NO, go to step A.5.

4. If the PCC uses Toxicall®, ask the PCC to run Report 57.
   a. To obtain occupational cases only: Cases should either have reason for the call (ExpReason) = 3 (occupational) OR exposure site (ExpSite) = 3 (workplace).
   b. To obtain all acute pesticide-related illnesses and injuries: Neither ExpReason nor ExpSite need to be specified.
   c. To calculate incidence rates, go to *B. Estimating the Total Population at Risk (denominator)*.

5. If the PCC does not use Toxicall® or if it cannot generate Report 57, determine if the PCC will provide a data set of all received calls.
   a. If YES, go to Step A.6.
   b. If NO, go to Step A.7.

6. If the PCC can provide a data set of all received calls, query the data set to identify cases that meet the following criteria:
   a. Exposure to an agent included in one of the pesticide generic categories
   (SubGenricCode) =
   Disinfectants
   0201008 disinfectant industrial cleaner
   0201055 bromine water/shock treatment
   0201056 chlorine water/shock treatment
   0042281 hypochlorite disinfectant: hypochlorite, non-bleach product
   0040280 phenol disinfectant: phenol (e.g., Lysol)
   0039282 pine oil disinfectant
   0077286 other/unknown disinfectant:

Fungicides (nonmedicinal)
0243566 carbamate fungicide
0201033 copper compound fungicide
0077564 mercurial fungicide
0077565 non-mercurial (inactive) fungicide
0253000 phthalimide fungicide
0254371 wood preservative
0077566 other/unknown (inactive) nonmedicinal fungicide
0201034 other nonmedicinal fungicide
0201035 unknown nonmedicinal fungicide

Fumigants
0201036 aluminum phosphide fumigant
0201037 metam sodium (fumigant, fungicide, or herbicide)
0201038 methyl bromide (fumigant, fungicide, or herbicide)
0201039 sulfuryl fluoride fumigant
0201040 other fumigant
0201041 unknown fumigant

Herbicides (includes algicides, defoliants, dessicants, plant growth regulators)
0201054 algicide
0254370 anti-algae paint
0243561 carbamate herbicide
0017000 2,4-d or 2,4,5-t (inactive)
0201042 chlorophenoxy herbicide
0049562 diquat
0201043 glyphosate
0049000 paraquat
0049561 paraquat/diquat combination
0077121 plant hormone
0213000 triazine herbicide
0215000 urea herbicide
0077561 other herbicide
0077567 unknown herbicide

Insecticides (includes insect growth regulators, molluscicides, nematicides)
0004562 arsenic pesticide
0062562 borate/boric acid pesticide
0070000 carbamate only (alone)
0070560 carbamate with other insecticide
0050000 chlorinated hydrocarbon only (alone)
0050560 chlorinated hydrocarbon with other insecticide
0201044 insect growth regulator

0172000 metaldehyde (molluscicide)
0208562 nicotine (excluding tobacco products)
0038000 organophosphate
0038560 organophosphate/carbamate combined
0038561 organophosphate/chlorinated hydrocarbon (inactive)
0038562 organophosphate/other insecticide
0038563 organophosphate/carbamate/chlorinated hydrocarbon (inactive)
0176000 piperonyl butoxide only (inactive)
0144000 piperonyl butoxide/pyrethrin (inactive) (without carbamate or o.p.)
0144001 pyrethrins only (inactive)
0201045 pyrethrin
0201046 pyrethroid
0145000 rotenone
0077568 veterinary insecticide (inactive) (for pets—flea collars, etc.)
0077562 other insecticide
0077569 unknown insecticide

Repellents
0201047 bird, dog, deer, or other mammal repellent
0201048 insect repellent with DEET
0201049 insect repellent without DEET
0218000 insect repellent: unknown (inactive)
0033000 naphthalene moth repellent
0050430 paradichlorobenzene moth repellent
0077431 other mothball or moth repellent
0077430 unknown mothball or moth repellent

Rodenticides
0174000 antu
0048563 anticoagulant: warfarin-type anticoagulant rodenticide
0048564 anticoagulant: long-acting, superwarfarin anticoagulant rodenticide
0244577 barium carbonate barium carbonate containing rodenticides
0201050 bromethalin
0201051 cholecalciferol rodenticide
0012563 cyanide rodenticide (excluding industrial or misc. chemical)
0162000 monofluoroacetate 1080/monofluoroacetate/smfa
0043000 strychnine rodenticide
0197000 vacor/pnu
0201052 zinc phosphide
0217000 thallium
0077563 other rodenticide
0077577 unknown rodenticide

b. Medical outcome (MedicalOutcome) is coded into one of the following values:
   201=minor effect
   202=moderate effect
   203=major effect
   204=death
   206=not followed, minimal clinical effects possible
   207=unable to follow, judged as a potentially toxic exposure

c. Request specific values for ExpReason and ExpSite, if needed.
   (1) To obtain occupational cases only: Cases should either have reason for the call (ExpReason) = 3 (occupational) OR exposure site (ExpSite) = 3 (workplace).
   (2) To obtain all acute pesticide-related illnesses and injuries: Neither ExpReason nor ExpSite need to be specified.

7. Using the case number, delete any duplicate cases.

8. Tally the total number of cases that meet the criteria.

9. If interested in calculating an incidence rate, go to *B. Estimating the Total Population at Risk (denominator)*.

10. If the PCC will not provide a data set:
    a. Ask the PCC to tally the number of cases that meet the criteria in A.6.a through A.6.d.
    b. If interested in calculating a rate, go to "B. Estimating the Total Population at Risk (denominator)."

B. Estimating the Total Population at Risk (denominator for rate calculations)
   1. Determine whether the rate is for acute occupational pesticide-related illness and injury, or for all acute pesticide-related illness and injury.
      a. If for acute occupational pesticide-related illness and injury, go to B.2.
      b. If for all acute pesticide-related illnesses and injuries, go to B.3.

   2. To obtain the denominator for an occupational case rate:
      a. Go to Current Population Statistics: http://www.bls.gov/opub/gp/laugp.htm.
      b. Select *Section II: Estimate for States*.
      c. Select *Table 12. Employment status of the civilian noninstitutional population by sex, age, race, and Hispanic origin*.
      d. Find your State from the first column.
      e. Read the *Total* row for your State and the 4th column—*Employment Number*. This is the *Number of Employer Persons 16 years of age or older* (in thousands). Multiply by 1000.
      f. Go to *C. Calculating the annual incidence rate*.

3. To obtain the denominator for the total population case rate:
   a. Use the US Census standard population. The most recent figures can be found at http://quickfacts.census.gov/qfd/index.html.
   b. After selecting your State, total population estimates will be provided.
   c. Go to *C. Calculating the annual incidence rate*.

C. Calculating the annual incidence rate
   1. Divide the numerator (A) by the denominator (B).
   2. Multiply this result by 100,000 to get the annual rate per 100,000 persons.

Appendix D
Case Definition for Acute Pesticide-Related Illness and Injury Cases Reportable to the National Public Health Surveillance System

Appendix D

Case Definition for Acute Pesticide-Related Illness and Injury Cases Reportable to the National Public Health Surveillance System

Contents:
D.1 Clinical Description
D.2 Laboratory Criteria for Diagnosis
D.3 Classification Criteria
D.4 Contacts for Additional Information
D.5 Frequently Asked Questions (FAQs)
D.6 Characteristic Signs and Symptoms for Several Pesticide Active Ingredients and Classes of Pesticides
D.7 Glossary of Terms

D.1 Clinical Description

This surveillance case definition refers to any acute adverse health effects resulting from exposure to a pesticide product (defined under the Federal Insecticide Fungicide and Rodenticide Act [FIFRA]) including health effects due to an unpleasant odor, injury from explosion of a product, inhalation of smoke from a burning product, and allergic reaction. Because public health agencies seek to limit all adverse effects from regulated pesticides, notification is needed even when the responsible ingredient is not the active ingredient.

A case is characterized by an acute onset of symptoms that are dependent on the formulation of the pesticide product and involve one or more of the following:

- Systemic signs or symptoms (including respiratory, gastrointestinal, allergic, and neurological signs/symptoms)
- Dermatologic lesions
- Ocular lesions

This case definition and classification system is designed to be flexible permitting classification of pesticide-related illnesses from all classes of pesticides. Consensus case definitions for specific classes of chemicals may be developed in the future.

A case will be classified as occupational if exposure occurs while at work (this includes working for compensation; in a family business, including a family farm; for pay at home; and as a volunteer emergency medical technician [EMT], firefighter, or law enforcement officer). All other cases will be classified as nonoccupational. All cases involving suicide or attempted suicide should be classified as nonoccupational.

A case is reportable to the national surveillance system when there is:

- Documentation of new adverse health effects that are temporally related to a documented pesticide exposure, *and*

- Consistent evidence of a causal relationship between the pesticide and the health effects based on the known toxicology of the pesticide from commonly available toxicology texts, government publications, information supplied by the manufacturer, or two or more case series or positive epidemiologic investigations, *or*

- Insufficient toxicologic information available to determine whether a causal relationship exists between the pesticide exposure and the health effects.

See the Classification Criteria section for a more detailed description of these criteria.

D.2 Laboratory Criteria for Diagnosis

If available, the following laboratory data can confirm exposure to a pesticide:

- Biological tests for the presence of, or toxic response to, the pesticide and/or its metabolite (in blood, urine, etc.)

  — Measurement of the pesticide and/or its metabolite(s) in the biological specimen

  — Measurement of a biochemical response to the pesticide in a biological specimen (e.g., cholinesterase levels)

- Environmental tests for the pesticide (e.g., foliage residue, analysis of suspect liquid)

- Pesticide detection on clothing or equipment used by the case subject

D.3 Classification Criteria

Reports received and investigated by State programs are scored on the three criteria provided below (criteria A, B, and C). Scores are either 1, 2, 3, or 4 and are assigned based on all available evidence. The classification matrix (Table D-1) provides the case classification categories and the criteria scores needed to place the case into a specific category. Definite, probable, possible, and suspicious cases (see the classification matrix) are reportable to the national surveillance system. Additional classification categories are provided for States that choose to track reports that do not fit the criteria for national reporting. (Frequently asked questions [FAQs] that provide additional clarification on the classification criteria and use of the classification matrix are provided in Section D.5. Section D.6 lists the characteristic signs and symptoms for several pesticide active ingredients and classes of pesticides.)

APPENDIX D

Table D-1. Case Classification Matrix

CLASSIFICATION CRITERIA	Definite Case	Probable Case	Possible Case	Suspicious Case	Unlikely Case	Insufficient Information	Not a Case: Asymptomatic[†]	Not a Case: Unrelated[‡]
A. Exposure	1	1	2	2	1 or 2	4	—	3
B. Health Effects	1	2	1	2	1 or 2	4	3	—
C. Causal Relationship	1	1	1	1	4	2	—	3

[*] Only reports meeting case classifications of Definite, Probable, Possible, and Suspicious are reportable to the NPHSS. Additional classification categories are provided for States that choose to track the reports that do not fit the national reporting criteria.

[†] The matrix does not indicate whether asymptomatic persons were exposed to pesticides although some States may choose to track the level of evidence of exposure for asymptomatic persons.

[‡] Unrelated = Illness determined to be caused by a condition other than pesticide exposure, as indicated by a 3 in the evidence of Exposure or Causal Relationship classification criteria.

A. DOCUMENTATION OF PESTICIDE EXPOSURE

1. Laboratory, clinical, or environmental evidence corroborates exposure *(at least one of the following must be satisfied to receive a score of 1)*:

   a. Analytical results from foliage residue, clothing residue, air, soil, water, or biologic samples.

   b. Observation of residue and/or contamination (including damage to plant material from herbicides) by a trained professional. (Note: a trained professional may be a plant pathologist, agricultural inspector, agricultural extension agent, industrial hygienist, or any other licensed or academically trained specialist with expertise in plant pathology and/or environmental effects of pesticides. A licensed pesticide applicator not directly involved with the application may also be considered a trained professional.)

   c. Biologic evidence of exposure (e.g., response to administration of an antidote such as 2-PAM, Vitamin K1, or repeated doses of atropine).

   d. Documentation by a licensed health care professional (HCP) of a characteristic eye injury or dermatologic effects at the site of direct exposure to a pesticide product known to produce such effects (these findings must be sufficient to satisfy criteria B.1 under "Documentation of Adverse Health Effect").

   e. Clinical description by a licensed HCP of two or more post-exposure health effects (at least one of which is a sign) characteristic for the pesticide as provided in Section D.6.

2. Evidence of exposure based solely on written or verbal report *(at least one of the following must be satisfied to receive a score of 2)*:

   a. Report by case

   b. Report by witness

c. Written records of application

   d. Observation of residue and/or contamination (including damage to plant material from herbicides) by other than a trained professional

   e. Other evidence suggesting that an exposure occurred

3. Strong evidence that no pesticide exposure occurred.

4. Insufficient data.

B. Documentation of Adverse Health Effect

1. Two or more new post-exposure abnormal signs and/or test/laboratory findings reported by a licensed HCP.

2. Two or more new post-exposure abnormal symptoms reported. When new post-exposure signs and test/laboratory findings are insufficient to satisfy a B1 score, they can be used in lieu of symptoms toward satisfying a B2 score.

3. No new post-exposure abnormal signs, symptoms, or test/laboratory findings reported.

4. Insufficient data (includes having only one new post-exposure abnormal sign, symptom, or test/laboratory finding).

C. Evidence Supporting a Causal Relationship Between Pesticide Exposure and Health Effects

1. Where the findings documented under the Health Effects criteria (criteria B) are:

   a. characteristic for the pesticide as provided in Section D.6, and the temporal relationship between exposure and health effects are plausible (the pesticide refers to the one classified under criteria A), *and/or*

   b. consistent with an exposure-health effect relationship based on the known toxicology (that is, exposure dose, symptoms, and temporal relationship) of the putative agent (that is, the agent classified under criteria A) from commonly available toxicology texts, government publications, information supplied by the manufacturer, or two or more case series or positive epidemiologic studies published in the peer-reviewed literature.

2. Evidence of exposure-health effect relationship is not present. This may be because the exposure dose was insufficient to produce the observed health effects. Alternatively, a temporal relationship does not exist (that is, health effects preceded the exposure or occurred too long after exposure). Finally, it may be because the constellation of health effects are not consistent based on the known toxicology of the putative agent from information in commonly available toxicology texts, government publications, information supplied by the manufacturer, or the peer-reviewed literature.

3. Definite evidence of nonpesticide causal agent.

4. Insufficient toxicologic information is available to determine causal relationship between exposure and health effects. (This includes circumstances where minimal human health effects data are available, or where there are less than two published case series or positive epidemiologic studies linking health effects to the particular pesticide product/ingredient or class of pesticides.)

D.4 Contacts for Additional Information

For information regarding acute occupational pesticide-related illness and injury, contact NIOSH at 1–800–35–NIOSH. For information about acute nonoccupational pesticide-related illness and injury, contact the National Center for Environmental Health (NCEH) at 404–639–2530. For information concerning regulation and use of pesticides, contact the US EPA, Office of Pesticide Programs, at 703–305–5336. The National Pesticide Information Center (NPIC) (1–800–858–7378) provides information about pesticides, acute pesticide-related illness and injury, and the toxicology and environmental chemistry of pesticides.

For more information about this case definition contact Geoffrey M. Calvert, M.D., M.P.H., at NIOSH (513–841–4448, e-mail jac6@cdc.gov).

Revised 11/29/04

D.5 Frequently Asked Questions (FAQs)

Question 1. *The terms* signs *and* symptoms *are used throughout the case definition. What is the difference between the two?*

Answer 1. *Signs* are objective findings that can be observed and described by a licensed HCP. Typically, this is the information found in the *physical exam* or *physical findings* section of a medical record or acute poisoning reporting form. These findings do not rely on the subjective reporting of sensations by the affected person. An objective, knowledgeable observer includes all licensed HCPs (e.g., medical doctor [MD], doctor of osteopathy [DO], physician's assistant [PA], registered nurse [RN], EMT, etc.).

Symptoms are any subjective evidence of a disease or a condition as perceived and reported by the patient. This includes reported changes from normal function, sensation, or appearance. This information is the *History* section of a medical record.

Question 2. *How should we classify the exposure when an affected person, their coworker, or family member indicates that they were "drenched" by pesticide spray?*

Answer 2. If no other corroborating evidence presented by an objective observer exists, the information meets criteria A2. If there is documentation by medical personnel, emergency responders (police, EMT, etc.), an employer, agency representative, or investigators that the person was observed to be drenched at the scene or treatment facility, this would be classified as meeting criteria A1b. However, it must be remembered that these observers must be objective and independent, and therefore they cannot be the affected person.

Question 3. How should an exposure be classified when a person has a dermal exposure that is difficult to document as a direct exposure? For example, a person handles an object contaminated with pesticides, then touches another part of the body with their possibly contaminated hand. The person then develops a dermal response at the site of hand contact.

Answer 3. If the person is confident that contact with the pesticide product definitely occurred, and the hand-to-body part contact occurred shortly afterward, and the dermal response is documented by a licensed HCP, code the exposure as A1d (documentation by a licensed HCP of a characteristic eye injury or dermatologic effects at the site of direct exposure to a pesticide product known to produce such effects). Code as A2 (evidence of exposure based solely on written or verbal report) if the dermal response is not documented by a licensed HCP. If the history is vague, or if contact may have been with a plant or product other than a pesticide, code as A4 (insufficient data).

Question 4. How do we interpret cholinesterase results when performing case classification?

Answer 4. Each State may choose to develop its own internal guidelines. The following very cursory discussion is provided to assist States in this process. *Cholinesterase depression* is defined as one or more of the following:

a. 30% depression from baseline (pre-exposure or 60 to 90 days post-exposure) red blood count cholinesterase level

b. 40% depression from baseline plasma cholinesterase level

c. Cholinesterase level below laboratory normal range

The level of depression may be determined by serial post-exposure testing if a baseline test is not available. (For example, testing 2 weeks and 4 weeks post exposure show a gradual increase in cholinesterase by percentages in 1 and 2 above, over the levels at initial testing.) A test that shows significant depression as described above should be considered evidence of exposure and ranked as meeting criteria A1c. It should also be considered evidence for a new post-exposure health effect and helps to meet the criteria for B1 (an additional post-exposure sign or test/laboratory finding would be needed to fully meet the criteria for B1). A test result that does not indicate depression should not be considered an indication that substantial exposure has not occurred. The timing of testing, laboratory variation, the wide normal range, and administration of praloxidime chloride (2PAM) prior to testing can all lead to negative results.

Question 5. Can the applicator who is directly affected by exposure or who has performed the application that is associated with health effects supply information that can be considered "evaluation by a trained professional" specified in criteria A1b?

Answer 5. No. Persons who are considered professional observers should be objective. An applicator who is the case cannot be considered an objective observer. Nor can an applicator be an objective observer when allegations or observations suggest a misapplication may have occurred. A trained, licensed applicator not directly involved with the case could be an observer under A1b. For example, a second applicator is called in to help evaluate damage to plants on the property, or to help alleviate odors in an office from an application by another applicator. This second person's observation can meet the requirements of a trained professional observer as specified in A1b.

Question 6. *What is the definition of* antidote *that should be used to evaluate exposure (A1c)?*

Answer 6. By *antidote*, we mean an agent that counteracts the effects of the pesticide. Two types of antidotes satisfy this definition: pharmacological antidotes and specific antidotes. Pharmacological antidotes counteract the pharmacological effects of the absorbed pesticide. Often, persons poisoned with pesticides have a high tolerance to repeated doses of pharmacological antidotes. For example, those poisoned with anticholinesterase pesticides have a high tolerance to atropine. As such, very high doses of atropine are often required to treat persons poisoned with anticholinesterase pesticides. Another pharmacological antidote is phenobarbital.

Specific antidotes interact directly with absorbed pesticide or some product of it to block the biochemical effect of the pesticide. Examples include pralidoxime chloride (2-PAM), vitamin K, and pesticide-specific monoclonal antibodies that are under development.

Antidotes are not the same as adjunct treatment that may help relieve symptoms or effects of the exposure in a less direct manner. This also does not include agents that prevent absorption of the ingested pesticide (e.g., activated charcoal).

Question 7. *How can we end up with a classification that is different from the clinical diagnosis in the medical record? Isn't that second guessing the physician's evaluation of the patient?*

Answer 7. The case classification scheme and the clinical diagnosis serve different purposes. The purpose of the case classification scheme is to serve surveillance and epidemiologic-related functions. The classification scheme provides objective guidelines for assessing the certainty of the evidence regarding exposure and health effects. In contrast, the purpose of the clinical diagnosis is to guide the immediate treatment course for the person. In addition, the clinician may use more intuitive and subjective criteria when making a diagnosis. Therefore, it is possible that the classification category may differ from the clinical diagnosis.

Question 8. *The classification scheme seems too stringent. By excluding persons who report only one symptom, we may be missing important cases. For example, a child with seizures after N, N-diethl-m-toluamide (DEET) exposure would be excluded. How can we address this?*

Answer 8. The classification scheme requires the presence of at least two post-exposure symptoms for a report to be considered a case. This may result in the exclusion of a very small number of actual pesticide-related illnesses or injuries. Most concerns about excluding cases due to this criterion can be alleviated by using structured protocols for obtaining medical histories from the person and/or HCP. If a single sign or symptom is reported, requesting more details will usually elicit additional signs or symptoms. Asking about commonly related symptoms as part of an interview is an acceptable practice.

For example, it is appropriate to ask about symptoms of nausea if a person reports vomiting, stomach cramping if diarrhea is reported, or loss of consciousness with seizure. This approach should help resolve concerns about the classification system resulting in false negatives.

Question 9. How do we assess signs and symptoms when a person has a pre-existing condition that may influence their physiologic response to an exposure?

Answer 9. Few studies have examined the effect of pre-existing disease on the toxicity of pesticides. We are not aware of any studies that found differences in signs and symptoms among pesticide-poisoned persons with pre-existing conditions. Therefore, if someone presents with an atypical set of symptoms for a particular pesticide, a score of C2 should be strongly considered under "evidence supporting a causal relationship between pesticide exposure and health effects."

However, it is possible that those with some pre-existing conditions will have reduced physiologic reserve. Therefore, these persons may manifest symptoms at a lower pesticide dose compared with a young, healthy person. Nonetheless, in these persons, the signs and symptoms should be characteristic of the particular pesticide, and the temporal relationship should be appropriate.

It is possible that pesticide exposure may exacerbate a pre-existing condition (e.g., organophosphate exposure can cause increased shortness of breath in exposed persons, including persons with chronic lung disease). However, the signs and symptoms that are present should be consistent with poisoning from the pesticide in question.

Question 10. How do we address a situation when the underlying condition may create a set of symptoms that are similar to the symptoms caused by the pesticide?

Answer 10. As has been stated previously, pesticide exposure may exacerbate a pre-existing condition. However, keep in mind that the signs and symptoms that are present should be consistent with poisoning from the pesticide in question. In addition, there should be an appropriate temporal relationship (that is, exposure preceded the health effect and the latency between exposure and effect is appropriate), and the pesticide exposure should be of sufficient dose.

Question 11. How do we determine whether the evidence for an exposure-health effect relationship is insufficient versus inconsistent?

Answer 11. When there is little literature on the health effects associated with a particular pesticide and none of it describes the health effects of interest, then the evidence for an exposure-health effect relationship is considered insufficient and a score of C4 is appropriate. However, if there are many references on the health effects associated with a particular pesticide, and none describe the health effects of interest, then the evidence for an exposure-health effect relationship is considered inconsistent and a score of C2 is appropriate.

Question 12. The term exposure dose *is used in Section C: "Evidence Supporting a Causal Relationship Between Pesticide Exposure and Health Effects." Often little information is available on dose. How should we interpret* dose?

Answer 12. The use of this term refers to whether the dose was sufficient to produce the observed health effects. Unfortunately, there is a paucity of data available on the minimum dose of a pesticide needed to produce health effects in humans. In addition, reaction to a pesticide exposure can vary across persons. It should be remembered that some persons may be much more sensitive to a pesticide and manifest health effects at a much lower dose compared with other persons. Other factors such as duration of exposure, use of protective equipment, amount of time between exposure and collection of the environmental sample, and the effect of intervening weather conditions on environmental samples and observations must be factored in when evaluating the actual exposure dose likely experienced by the person. When available, the peer-reviewed literature should be examined for guidance. The judgment of colleagues in the State Department of Agriculture may also be helpful.

When dealing with self-reports, qualitative information about exposure dose can be obtained. For example, information can be obtained about proximity to the source of exposure, duration of exposure, did health effects manifest in others who were exposed, etc. Assessing this information may require experience and the assistance of other knowledgeable colleagues.

Question 13. *Often we learn that a person was exposed to a particular functional class of pesticides (e.g., insecticide, herbicide, etc.), but we cannot determine the name of the product or the active ingredient. Should an exposure score of A2=written or verbal report or A4=insufficient data be assigned?*

Answer 13. When only the pesticide class is known, a score of A4=insufficient data must be assigned. This is because the pesticides within a particular class can vary widely in toxicity. Therefore, it would be impossible to determine if any observed health effects are consistent and/or characteristic with the pesticide exposure. However, if the chemical class of the pesticide is known (e.g., organophosphate or carbamate), but the specific pesticide product or active ingredient is unknown, a score of A1 or A2 can be considered. This is because pesticides within a specific chemical class can produce similar health effects (see Section D.6).

Question 14. *Can documentation or a clinical description "by a licensed HCP" as specified in criteria A1d, A1e, and B1 be provided by the licensed HCP who is directly affected by exposure (please note that this is similar to question Question 5)?*

Answer 14. No. Persons who are considered professional observers should be objective. An HCP who is the case cannot be considered an objective observer. A licensed HCP not directly involved in the exposure event would meet the criteria under A1d, A1e, and B1.

D.6. CHARACTERISTIC SIGNS AND SYMPTOMS FOR SEVERAL PESTICIDE ACTIVE INGREDIENTS AND CLASSES OF PESTICIDES

Pesticide	Signs and Symptoms
Acrolein	Conjunctivitis (irritation of mucous membranes, tearing) Skin irritation, rash, blistering, or erosion (without sensitization) Pulmonary edema Tearing Upper respiratory tract irritation: rhinitis, scratchy throat, cough
Acrylonitrile	Seizures/convulsions (tonic-clonic), sometimes leading to coma Upper respiratory tract irritation: rhinitis, scratchy throat, cough
Aminopyridine	Behavioral-mood disturbances (confusion, excitement, mania, disorientation, emotional lability) Salivation Sweating (diaphoresis) Thirst
ANTU	Dyspnea Upper respiratory tract irritation: rhinitis, scratchy throat, cough
Arsenicals (inorganic)	Anemia Abdominal pain Behavioral-mood disturbances (confusion, excitement, mania, disorientation, emotional lability) Bloody diarrhea Keratoses, brown discoloration Kidney (proteinuria, hematuria, sometimes leading to oliguria, acute renal failure with azotemia Leukopenia, thrombocytopenia Metallic taste in mouth Paralysis, paresis (muscle weakness) Paresthesia of extremities Runny nose Stomatitis Thirst
Arsine	Anemia Chills Hemoglobinuria Hemolysis Hyperkalemia Kidney (proteinuria, hematuria, sometimes leading to oliguria, acute renal failure with azotemia)

Pesticide	Signs and Symptoms
Borate	Abdominal pain
	Beefy red palms, soles
	Diarrhea
	Hypotension, shock
	Kidney (proteinuria, hematuria, sometimes leading to oliguria, acute renal failure with azotemia)
	Nervous system depression (stupor, coma, respiratory failure, often without seizures/convulsions)
	Tremor
Cadmium compounds	Abdominal pain
	Conjunctivitis (irritation of mucous membranes, tearing)
	Cyanosis
	Diarrhea
	Dyspnea
	Pulmonary consolidation
	Pulmonary edema
	Salivation
	Skin irritation, rash, blistering, or erosion (without sensitization)
	Upper respiratory tract irritation: rhinitis, scratchy throat, cough
Carbamate insecticides	Abdominal pain
	Anorexia
	Bradycardia (sometimes to asystole)
	Diarrhea
	Diplopia
	Dyspnea
	Incoordination (including ataxia)
	Miosis
	Muscle twitching
	Nervous system depression (stupor, coma, respiratory failure, often without seizures/convulsions)
	Paralysis, paresis (muscle weakness)
	Runny nose
	Salivation
	Sweating (diaphoresis)
	Tearing
	Tremor

Pesticide	Signs and Symptoms
Carbon disulfide	Behavioral-mood disturbances (confusion, excitement, mania, disorientation, emotional lability) Breath odor of rotten cabbage Incoordination (including ataxia) Paresthesia of extremities Seizures/convulsions (tonic-clonic), sometimes leading to coma
Carbon tetrachloride	Jaundice Liver enlargement Liver enzymes elevated (lactate dehydrogenase [LDH], alanine amiotransferase [ALT], aspartate transaminase [AST], alkaline phosphatase)
Cationic detergents	Skin irritation, rash, blistering, or erosion (without sensitization) Pulmonary edema
Chlordimeform	Anorexia Hot sensations Kidney (dysuria, hematuria, pyuria) Skin irritation, rash, blistering, or erosion (without sensitization) Sweet taste in mouth
Chlorhexidine	Contact dermatitis Urticaria
Chloroform	Jaundice Liver enlargement Liver enzymes elevated (LDH, ALT, AST, alkaline phosphatase)
Chloropicrin	Conjunctivitis (irritation of mucous membranes, tearing) Dyspnea Tearing Upper respiratory tract irritation: rhinitis, scratchy throat, cough
Cholecalciferol	Anorexia Hypercalcemia Polyuria Thirst
Copper compounds	Abdominal pain Conjunctivitis (irritation of mucous membranes, tearing) Hypotension, shock Kidney (proteinuria, hematuria, sometimes leading to oliguria, acute renal failure with azotemia)

Pesticide	Signs and Symptoms
Copper compounds (*continued*)	Liver enlargement Skin irritation, rash, blistering, or erosion (without sensitization) Stomatitis
Coumarins	Bloody diarrhea Ecchymoses Hypoprothrombinemia
Creosote	Contact dermatitis Hypothermia Methemoglobinemia Pallor Pulmonary edema Seizures/convulsions (tonic-clonic), sometimes leading to coma Smoky urine
Crimidine	Cyanosis Seizures/convulsions (tonic-clonic), sometimes leading to coma
Cyanamide	Dyspnea Hypotension, shock Skin flushing Tachycardia
Cyanide	Behavioral-mood disturbances (confusion, excitement, mania, disorientation, emotional lability) Bradycardia (sometimes to asystole) Breath odor of bitter almonds Dilated pupils Salivation Seizures/convulsions (tonic-clonic), sometimes leading to coma Unreactive pupils
DEET	Contact dermatitis Seizures/convulsions (tonic-clonic), sometimes leading to coma Urticaria
Dibromochloropropane	Low sperm count Skin irritation, rash, blistering, or erosion (without sensitization)

Pesticide	Signs and Symptoms
Diquat	Abdominal pain
	Behavioral-mood disturbances (confusion, excitement, mania, disorientation, emotional lability)
	Bloody diarrhea
	Conjunctivitis (irritation of mucous membranes, tearing)
	Ileus
	Kidney (proteinuria, hematuria, sometimes leading to oliguria, acute renal failure with azotemia)
	Nervous system depression (stupor, coma, respiratory failure, often without seizures/convulsions)
	Skin irritation, rash, blistering, or erosion (without sensitization)
	Stomatitis
Endothall	Bloody diarrhea
	Conjunctivitis (irritation of mucous membranes, tearing)
	Hypotension, shock
	Seizures/convulsions (tonic-clonic), sometimes leading to coma
	Skin irritation, rash, blistering, or erosion (without sensitization)
Ethylene dibromide	Kidney (proteinuria, hematuria, sometimes leading to oliguria, acute renal failure with azotemia)
	Skin irritation, rash, blistering, or erosion (without sensitization)
	Pulmonary edema
	Upper respiratory tract irritation: rhinitis, scratchy throat, cough
Ethylene oxide	Cardiac arrhythmias
	Conjunctivitis (irritation of mucous membranes, tearing)
	Dermal sensitization
	Pulmonary edema
	Skin irritation, rash, blistering, or erosion (without sensitization)
Fluoride	Abdominal pain
	Bloody diarrhea
	Dilated pupils
	Hypocalcemia
	Seizures/convulsions (tonic-clonic), sometimes leading to coma
	Tetany, carpopedal spasms
Formaldehyde	Conjunctivitis (irritation of mucous membranes, tearing)
	Skin irritation, rash, blistering, or erosion (without sensitization)
	Upper respiratory tract irritation: rhinitis, scratchy throat, cough

Pesticide	Signs and Symptoms
Fumigants (halocarbon)	Cardiac arrhythmias Incoordination (including ataxia)
Hexachlorobenzene	Anorexia Porphyrinuria (wine-red urine)
Hexachlorophene	Contact dermatitis Seizures/convulsions (tonic-clonic), sometimes leading to coma Skin irritation, rash, blistering, or erosion (without sensitization)
Indandiones	Bloody diarrhea Ecchymoses Hypoprothrombinemia
Mercury (organic)	Behavioral-mood disturbances (confusion, excitement, mania, disorientation, emotional lability) Constricted eye fields Hearing loss Metallic taste in mouth Paresthesia of extremities Tremor
Metaldehyde	Abdominal pain Seizures/convulsions (tonic-clonic), sometimes leading to coma Tremor
Metam sodium	Conjunctivitis (irritation of mucous membranes, tearing) Skin irritation, rash, blistering, or erosion (without sensitization)
Methyl bromide	Behavioral-mood disturbances (confusion, excitement, mania, disorientation, emotional lability) Dyspnea Conjunctivitis (irritation of mucous membranes, tearing) Pulmonary consolidation Pulmonary edema Skin irritation, rash, blistering, or erosion (without sensitization)
Naphthalene	Anemia Conjunctivitis (irritation of mucous membranes, tearing) Hemoglobinuria Hemolysis Hyperkalemia

Pesticide	Signs and Symptoms
Naphthalene (*continued*)	Kidney (proteinuria, hematuria, sometimes leading to oliguria, acute renal failure with azotemia) Sweating (diaphoresis) Upper respiratory tract irritation: rhinitis, scratchy throat, cough
Nicotine	Abdominal pain Anorexia Behavioral-mood disturbances (confusion, excitement, mania, disorientation, emotional lability) Cardiac arrhythmias Diarrhea Cyanosis Diplopia Dyspnea Hypertension (early in poisoning) Incoordination (including ataxia) Muscle twitching Paralysis, paresis (muscle weakness) Salivation Seizures/convulsions (tonic-clonic), sometimes leading to coma Sweating (diaphoresis) Tremor
Nitrophenols	Behavioral-mood disturbances (confusion, excitement, mania, disorientation, emotional lability) Fever Hot sensations Kidney (proteinuria, hematuria, sometimes leading to oliguria, acute renal failure with azotemia) Skin flushing Sweating (diaphoresis) Tachycardia Thirst Yellow stain on skin Yellow sclera
Organochlorines	Cyanosis Pallor Paresthesia (chiefly facial, transitory) Seizures/convulsions (tonic-clonic), sometimes leading to coma

Pesticide	Signs and Symptoms
Organophosphates	Abdominal pain
	Acetylcholinesterase depression (RBC and/or plasma)
	Anorexia
	Bradycardia (sometimes to asystole)
	Diarrhea
	Diplopia
	Dyspnea
	Incoordination (including ataxia)
	Miosis
	Muscle twitching
	Nervous system depression (stupor, coma, respiratory failure, often without seizures/convulsions)
	Paralysis, paresis (muscle weakness)
	Paresthesia (chiefly facial, transitory)
	Runny nose
	Salivation
	Sweating (diaphoresis)
	Tearing
	Tremor
Organotin compounds	Abdominal pain
	Behavioral-mood disturbances (confusion, excitement, mania, disorientation, emotional lability)
	Conjunctivitis (irritation of mucous membranes, tearing)
	Skin irritation, rash, blistering, or erosion (without sensitization)
Paraquat	Abdominal pain
	Bloody diarrhea
	Conjunctivitis (irritation of mucous membranes, tearing)
	Contact dermatitis
	Cyanosis
	Dyspnea
	Jaundice
	Keratitis
	Kidney (proteinuria, hematuria, sometimes leading to oliguria, acute renal failure with azotemia)
	Myalgia
	Pulmonary consolidation
	Skin irritation, rash, blistering, or erosion (without sensitization)
	Stomatitis
	Upper respiratory tract irritation: rhinitis, scratchy throat, cough

Pesticide	Signs and Symptoms
Pentachlorophenol	Anorexia Contact dermatitis Dyspnea Fever Kidney (proteinuria, hematuria, sometimes leading to oliguria, acute renal failure with azotemia) Sweating (diaphoresis) Tachycardia Thirst Urticaria
Phosphorus	Abdominal pain Breath odor of garlic Hypotension, shock Jaundice Pulmonary edema Skin irritation, rash, blistering, or erosion (without sensitization) Tetany, carpopedal spasms Thirst
Phosphides	Abdominal pain Breath odor of garlic Hypotension, shock Jaundice Paresthesia (chiefly facial, transitory) Pulmonary edema Tetany, carpopedal spasms Thirst
Phosphine	Breath odor of garlic Chills Hypotension, shock Jaundice Liver enlargement Liver enzymes elevated (LDH, ALT, AST, alkaline phosphatase) Pulmonary edema Seizures/convulsions (tonic-clonic), sometimes leading to coma Thirst
Povidone-iodine	Cardiac arrhythmias Seizures/convulsions (tonic-clonic), sometimes leading to coma

Pesticide	Signs and Symptoms
Propargite	Dermal sensitization Skin irritation, rash, blistering, or erosion (without sensitization)
Pyriminil	Behavioral-mood disturbances (confusion, excitement, mania, disorientation, emotional lability) Breath odor of peanuts Cardiac arrhythmias Constipation Glucosuria Hyperglycemia (elevated serum glucose) Ketoacidosis Ketonuria Paresthesia of extremities Urinary retention
Pyrethrins	Contact dermatitis Runny nose
Pyrethroids	Diarrhea Pulmonary edema
Sabadilla	Cardiac arrhythmias Sneezing
Sodium chlorate	Anemia Cardiac arrhythmias Cyanosis Hemoglobinuria Hemolysis Hyperkalemia Hypotension, shock Jaundice Kidney (proteinuria, hematuria, sometimes leading to oliguria, acute renal failure with azotemia)) Liver enlargement Methemoglobinemia Seizures/convulsions (tonic-clonic), sometimes leading to coma Skin irritation, rash, blistering, or erosion (without sensitization)

Pesticide	Signs and Symptoms
Sodium fluoride	Cardiac arrhythmias Hypotension, shock Kidney (proteinuria, hematuria, sometimes leading to oliguria, acute renal failure with azotemia) Pallor Nervous system depression (stupor, coma, respiratory failure, often without seizures/convulsions) Salivation Salty, soapy taste in mouth Thirst
Sodium fluoroacetate	Behavioral-mood disturbances (confusion, excitement, mania, disorientation, emotional lability) Cardiac arrhythmias Cyanosis Paresthesia of extremities Seizures/convulsions (tonic-clonic), sometimes leading to coma
Strychnine	Cyanosis Seizures/convulsions (tonic-clonic), sometimes leading to coma
Sulfur	Breath odor of rotten eggs Diarrhea Skin irritation, rash, blistering, or erosion (without sensitization)
Sulfur dioxide	Conjunctivitis (irritation of mucous membranes, tearing) Dyspnea Pulmonary edema Upper respiratory tract irritation: rhinitis, scratchy throat, cough
Sulfuryl fluoride	Dyspnea Kidney (proteinuria, hematuria, sometimes leading to oliguria, acute renal failure with azotemia) Muscle twitching Upper respiratory tract irritation: rhinitis, scratchy throat, cough
Thallium	Abdominal pain Behavioral-mood disturbances (confusion, excitement, mania, disorientation, emotional lability) Bloody diarrhea Cardiac arrhythmias (ventricular) Hypertension (early in poisoning)

Pesticide	Signs and Symptoms
Thallium	Hypotension, shock Ileus Incoordination (including ataxia) Loss of hair Paresthesia of extremities Ptosis Seizures/Convulsions (tonic-clonic), sometimes leading to coma Tremor
Thiram	Alcohol intolerance Contact dermatitis Diarrhea Skin irritation, rash, blistering, or erosion (without sensitization)
Veratrum alkaloid	(See sabadilla)

(Adapted from Morgan DP. Recognition and management of pesticide poisonings. 4th ed. Washington, DC: U.S. EPA; 1989; and Reigart JR, Roberts JR. Recognition and management of pesticide poisonings. 5th ed. Washington, DC: U.S. EPA; 1999.)

D.7 Glossary of Medical Terms

Anorexia	diminished appetite
Bradycardia	slow heart rate (generally less than 60 beats per minute)
Carpopedal spasms	spasm of the hands and/or feet
Conjunctivitis	inflammation of the conjunctiva (the mucous membrane covering the surface of the eye)
Cyanosis	a dark blueish or purplish coloration of the skin and mucous membranes
Diaphoresis	sweating, perspiration
Dyspnea	shortness of breath
Ecchymoses	bruises of the skin larger than 3mm in diameter
Glucosuria	presence of glucose in the urine
Hemoglobinuria	presence of hemoglobin in the urine
Hemolysis	destruction of red blood cells
Hypercalcemia	increased calcium in the blood
Hyperkalemia	increased potassium in the blood
Hypertension	increased blood pressure
Hypoprothrombinemia	low levels of prothrombin in the blood
Hypothermia	decreased body temperature (significantly below 98.6 F)
Ileus	obstruction of the bowel
Keratoses	a hard, thick circumscribed skin lesion (characterized by overgrowth of the horny layer)
Ketoacidosis	an increase in the pH of the blood caused by the enhanced production of ketones
Ketouria	presence of ketones in the urine
Leukopenia	decreased number of white blood cells in the blood
Methemoglobinemia	the presence of methemoglobin in the blood

Appendix D

Miosis	pinpoint pupils
Myalgia	muscular pain
Paresis	muscle weakness
Paresthesia	an abnormal sensation such as of burning, pricking, tingling or tickling
Polyuria	increased production of urine resulting in increased frequency of urination
Porphyrinuria	increased porphyrins in the urine manifesting as wine-red urine
Ptosis	a sinking down of the eyelid
Pulmonary consolidation	an infiltrate in the lung observed on a chest x-ray
Rhinitis	inflammation of the nasal mucous membranes
Stomatitis	inflammation of the mucous membranes of the mouth
Tachycardia	rapid heart rate (generally greater than 100 beats per minute)
Tetany	a clinical neurological syndrome characterized by muscle twitches, cramps, carpopedal spasm, and when severe, laryngospasm and seizures
Thrombocytopenia	decreased number of platelets in the blood

(January 31, 2000)

Appendix E
Severity Index for Use in State-Based Surveillance of Acute Pesticide-Related Illness and Injury

Appendix E

Severity Index for Use in State-Based Surveillance of Acute Pesticide-Related Illness and Injury

Purpose

The purpose of the severity index is to provide simple, standardized criteria for assigning severity to cases of acute pesticide-related illness and injury.

Rationale

It is important to assign a severity category to each case of acute pesticide-related illness and injury. An understanding of illness severity will be useful to evaluate the morbidity of acute pesticide-related illness and injury, to assess its impact on society, and to assist the targeting of limited intervention/prevention resources toward the most pressing pesticide problems.

Description

This severity index is based upon existing systems for ranking severity of poisonings, including pesticide illness [AAPCC 1992; Washington Department of Health 1999; EPA 1998; Persson et al. 1998]. It takes into account the following: signs and symptoms, whether medical care was sought, whether the individual was hospitalized, and whether lost time from work or usual activities occurred. Severity should be assigned only to acute pesticide-related illnesses or injuries classified as definite, probable, possible, or suspicious. As such, this severity index should be used in conjunction with the *Case Definition for Acute Pesticide-Related Illness and Injury Cases Reportable to the National Public Health Surveillance System* [NIOSH 2004].

Figure 1 is the flow diagram that should be used as a guide for assigning severity. The figure often refers to "the Table." This is Table 1, a listing of signs and symptoms that correspond to the different severity categories. Many of the signs and symptoms in the table are included in *Standardized Variables for Pesticide Poisoning Surveillance* [NIOSH 2000]. When using the table, only signs and symptoms related to the pertinent acute pesticide-related illness or injury should be considered (that is, only consider those signs and symptoms used to classify the acute pesticide-related illness and injury as definite, probable, possible, or suspicious).

The list of signs and symptoms provided in the table is not comprehensive, but instead provides examples to assist in assessing severity. In addition, a given health effect may appear in more than one of the table's severity columns. In such instances, the health effect observed as a sign (that is, a health effect observed and described by a licensed HCP) will be considered as having greater severity compared to the health effect reported as a symptom (that is, a health effect perceived and reported by the patient but not observed by a licensed HCP).

This severity index provides standardized criteria to ensure inter-rater uniformity in assigning severity. However, we recognize that this severity index cannot address all conceivable clinical situations. Therefore, it is not realistic to insist on strict adherence to these criteria. The user must be flexible when using this severity index, given that the user will not infrequently need to employ judgment and experience when assigning severity.

A brief description of each of the four severity categories follows.

S-1 Death

This category describes a human fatality resulting from exposure to one or more pesticides.

S-2 High severity illness or injury

The illness or injury is severe enough to be considered life threatening and typically requires treatment. This level of effect commonly involves hospitalization to prevent death. Signs and symptoms include, but are not limited to, coma, cardiac arrest, renal failure, and/or respiratory depression. The individual sustains substantial loss of time (> 5 days) from regular work (this can include assignment to limited/light work duties) or normal activities (if not employed). This level of severity might include the need for continued health care following the exposure event, prolonged time off work, and limitations or modification of work or normal activities. The individual may sustain permanent functional impairment.

S-3 Moderate severity illness or injury

This category includes cases of less severe illness or injury often involving systemic manifestations. Generally, treatment was provided. The individual is able to return to normal functioning without any residual disability. Usually, less time is lost from work or normal activities (3 to 5 days), compared with those with severe illness or injury. No residual impairment is present (although effects may be persistent).

S-4 Low severity illness or injury

This is the category of lowest severity. It is often manifested by skin, eye, or upper respiratory irritation and may also include fever, headache, fatigue, or dizziness. Typically the illness or injury resolves without treatment. There is minimal lost time (< 3 days) from work or normal activities.

References

AAPCC [1992]. Toxic Exposure Surveillance System (TESS) Manual. Washington, DC: American Association of Poison Control Centers.

EPA [1998]. Expanded explanation for the new FIFRA 6(a)(2)159.814 (5)(i)(A-E) and (5)(ii)(A E) exposure severity categories. Washington, DC: U.S. Environmental Protection Agency.

NIOSH [2004]. Case definition for acute pesticide-related illness and injury cases reportable to the national public health surveillance system. Cincinnati, OH: U.S. Department of Health and Human Services, Centers for Disease Control and Prevention, National Institute for Occupational Safety and Health, 2000. [http://www.cdc.gov/niosh/topics/pesticides/]

NIOSH [2000]. Standardized variables for state surveillance of pesticide-related illness and injury. Cincinnati, OH: U.S. Department of Health and Human Services, Centers for Disease Control and Prevention, National Institute for Occupational Safety and Health, 2000. Unpublished.

Persson HE, Sjoberg GK, Haines JA, de Garbino JP [1998]. Poisoning severity score. Grading of acute poisoning. Clin Toxicol 36:205B213.

WSDOH [1999]. 1998 Annual Report, Pesticide Incident Reporting and Tracking Review Panel. Olympia, WA: Washington State Department of Health (WSDOH), Office of Environmental Health and Safety.

Table 1. Signs and symptoms by severity category. (Modeled after Persson et al. 1998 and includes SPIDER database elements.)

ORGAN SYSTEM	SEVERITY CATEGORY AND CODE			
	DEATH 1	HIGH 2	MODERATE 3	LOW 4
	Fatal	Severe or life-threatening signs	Pronounced or prolonged signs or symptoms	Mild, transient, and spontaneously resolving symptoms
Gastrointestinal system		• Massive hemorrhage/perforation of gut	• Diarrhea (GI4, sign only) • Melena (GI7) • Vomiting (GI6, sign only)	• Abdominal pain, cramping (GI1) • Anorexia (GI2) • Constipation (GI3) • Diarrhea (GI4, symptom) • Nausea (GI5) • Vomiting (GI6, symptom)
Respiratory system		• Cyanosis (RESP 2) + Respiratory depression (RESP 7) • Pulmonary edema (RESP6) • Respiratory arrest	• Abnormal pulmonary x-ray • Pleuritic chest pain/pain on deep breathing (RESP8) • Respiratory depression (RESP7) • Wheezing (RESP9) • Dyspnea, shortness of breath (RESP4, sign only)	• Cough (RESP1) • Upper respiratory pain, irritation (RESP3) • Dyspnea, shortness of breath (RESP4, symptom)
Nervous system		• Coma (NS3) • Paralysis, generalized (NS10) • Seizure (NS5, sign only)	• Confusion (NS4) • Hallucinations (NS99 Other) • Miosis with blurred vision (NS14) • Seizure (NS5, symptom) • Ataxia (NS1, sign only) • Slurred speech (NS12) • Syncope (fainting) (NS17) • Peripheral neuropathy (NS11, sign only)	• Hyperactivity (NS2) • Headache (NS7) • Profuse sweating (NS13) • Dizziness (NS15) • Ataxia (NS1, symptom) • Peripheral neuropathy (NS11, symptom)
Cardiovascular system		• Bradycardia/ heart rate <40 for adults, < 60 infants and children, <80 neonates (CV1) • Tachycardia/ heart rate >180 for adults, >190 infants/children, >200 in neonates (CV4) • Cardiac arrest (CV2)	• Bradycardia / heart rate 40-50 in adults, 60-80 in infants/children, 80-90 in neonates (CV1) • Tachycardia / heart rate=140-180 in adults, 160-190 infants/children, 160-200 in neonates (CV4) • Chest Pain (CV7) + Hyperventilation, Tachypnea (RESP5) • Conduction disturbance (CV3) • Hypertension (CV6) • Hypotension (CV5)	

APPENDIX E

Table 1 (Continued). Signs and symptoms by severity category. (Modeled after Persson et al. 1998 and includes SPIDER database elements.)

ORGAN SYSTEM	SEVERITY CATEGORY AND CODE			
	DEATH 1	HIGH 2	MODERATE 3	LOW 4
	Fatal	Severe or life-threatening signs	Pronounced or prolonged signs or symptoms	Mild, transient, and spontaneously resolving symptoms
Metabolism		• Acid Base disturbance (pH< 7.15 or >7.7)	• Acid Base disturbance (pH = 7.15-7.24 or 7.60-7.69) • Elevated anion gap (MISC4)	• Fever (MISC1)
Renal system		• Anuria (GU2) • Renal failure	• Hematuria (GU3) • Oliguria (GU2) • Proteinuria (GU4)	• Polyuria (GU1)
Muscular system		• Muscle rigidity (NS9) + elevated urinary myoglobin + elevated creatinine	• Fasciculations (NS6) • Muscle rigidity (NS9) • Muscle weakness (NS8, sign only)	• Muscle weakness (NS8, symptom) • Muscle pain (NS16)
Local effects on skin		• Burns, second degree (involving >50% of body surface area) • Burns, third degree (involving >2% of body surface area)	• Bullae (DERM1) • Burns, second degree (involving <50% of body surface area) • Burns, third degree (involving <2% of body surface area)	• Skin Edema/Swelling, Erythema, Rash, Irritation/Pain, Pruritus (DERM3 - 7) • Hives/Urticaria
Local effects on eye		• Corneal ulcer/perforation	• Corneal abrasion (EYE3) • Ocular burn (EYE2)	• Lacrimation (EYE4) • Mydriasis (EYE6) • Miosis (EYE1) • Ocular pain/irritation/inflammation (diagnosis of conjunctivitis) (EYE5)
Other effects				• Fatigue (MISC5) • Malaise (MISC6)

Rev. 11/27/01

Appendix E

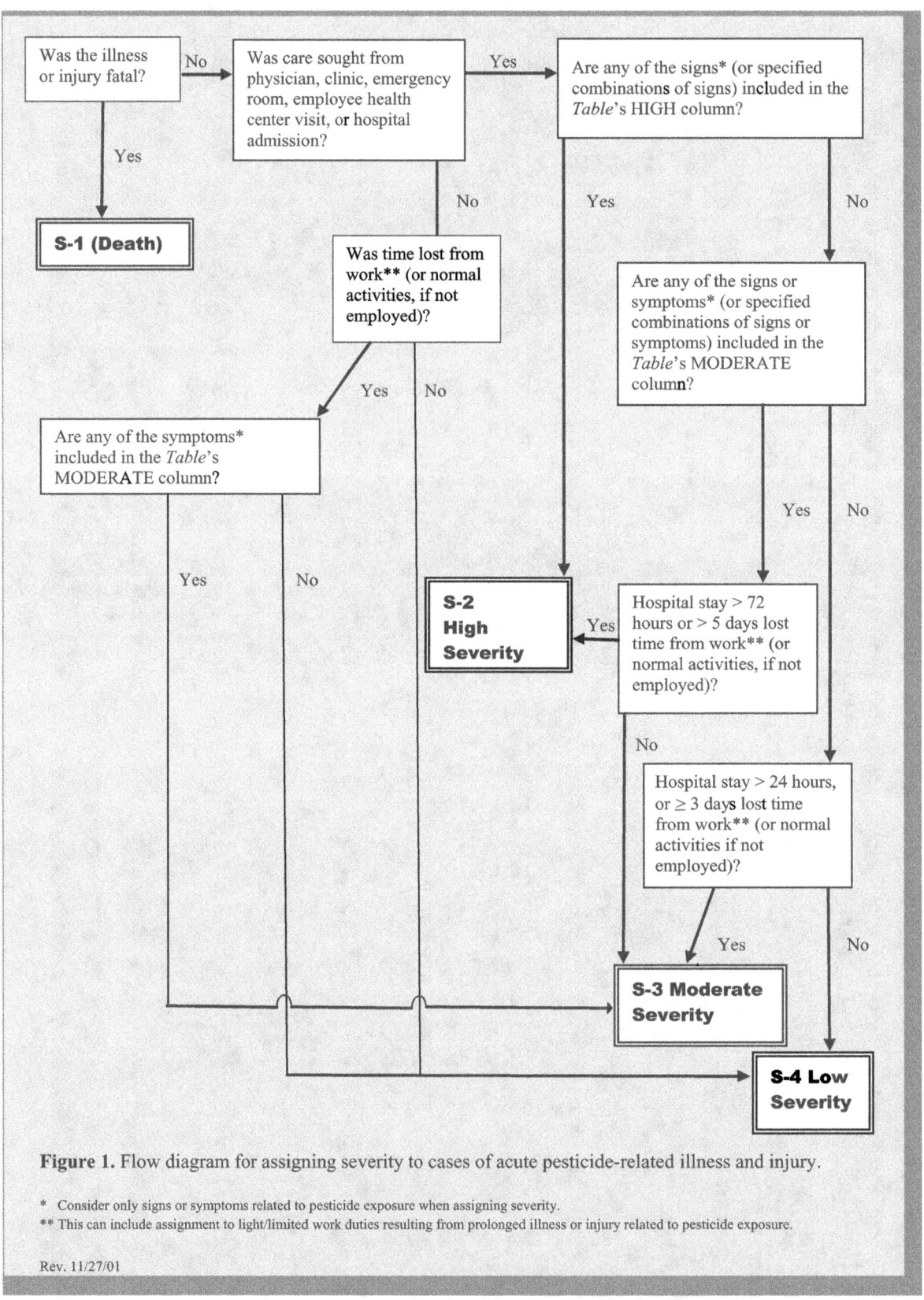

Figure 1. Flow diagram for assigning severity to cases of acute pesticide-related illness and injury.

* Consider only signs or symptoms related to pesticide exposure when assigning severity.
** This can include assignment to light/limited work duties resulting from prolonged illness or injury related to pesticide exposure.

Rev. 11/27/01

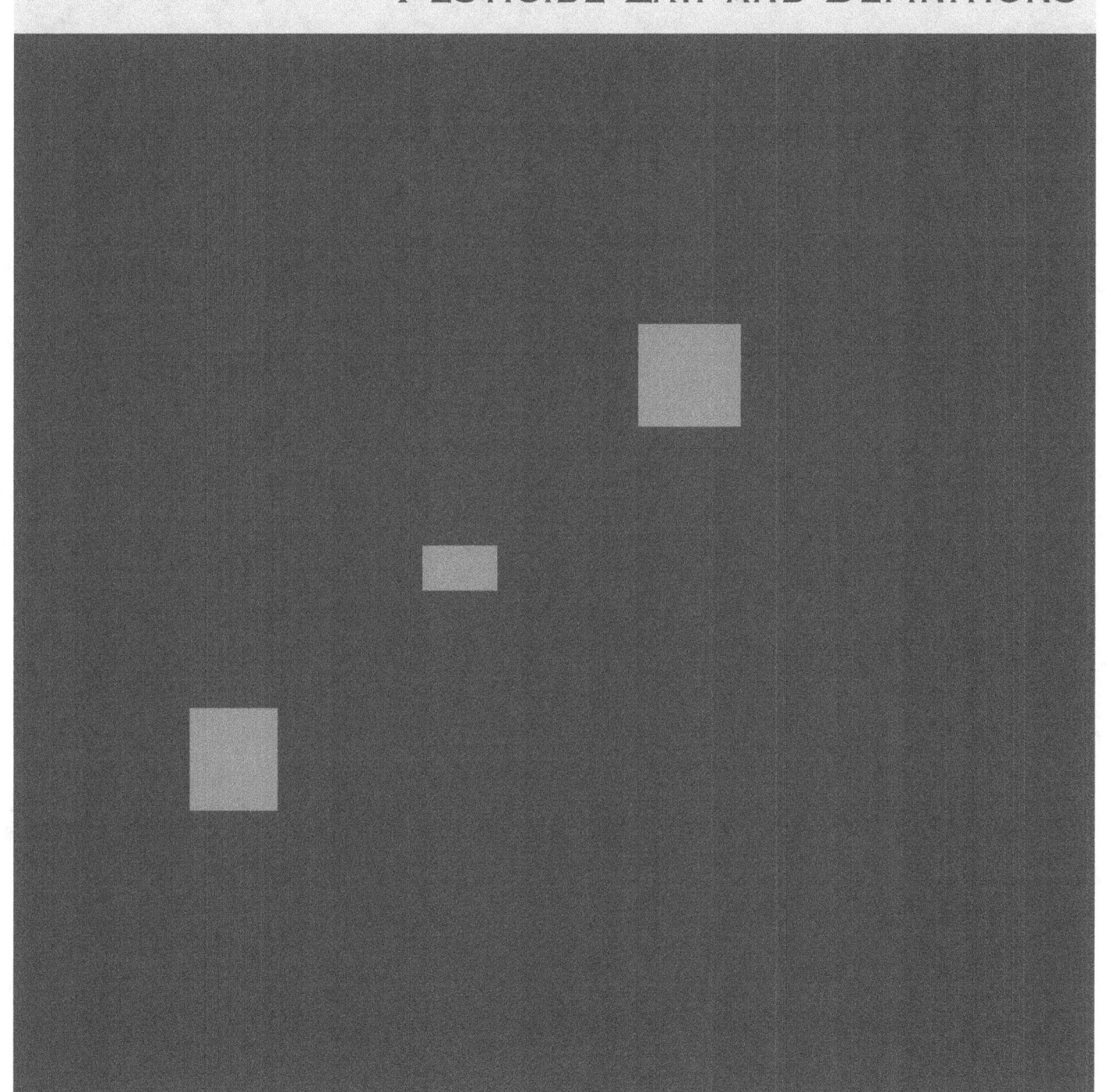

Appendix F
Pesticide Law and Definitions

Appendix F

Pesticide Law and Definitions

This appendix briefly outlines some of the Federal laws that regulate pesticides in the United States. It does not include every law that pertains to pesticides but touches on those most pertinent for surveillance of acute human illness and injury. The U.S. Environmental Protection Agency (EPA) is responsible for the regulation of pesticide products marketed in the United States. Each State will have its own set of statutes and administrative rules that address pesticides and reflect these Federal laws.

Contents:
F.1 Federal Insecticide Fungicide and Rodenticide Act (FIFRA)
F.2 Federal Food, Drugs, and Cosmetic Act (FFDCA)
F.3 Food Quality Protection Act (FQPA)
F.4 Safe Drinking Water Act (SDWA)
F.5 Occupational Safety and Health Act (OSH Act)

F.1 Federal Insecticide, Fungicide, and Rodenticide Act (FIFRA)

FIFRA was first passed in 1947. It was primarily a consumer protection law aimed at protecting pesticide users from products that did not contain active ingredients or sufficient active ingredients to be effective. FIFRA underwent extensive amendments by Congress in 1972 that required comprehensive testing of both old and new products. This required the review of 600 active ingredients and 50,000 registered pesticide products. These amendments shifted the emphasis of FIFRA to protecting human health and the environment from unreasonable adverse risks associated with the use of pesticides. This was intended by Congress to include protections for farmers, farmworkers, and others who were in contact with pesticides. The regulatory approach required an evaluation of risks and benefits from the use of pesticide products. The data requirements issued in 1975 were quite extensive. They required manufacturers to provide information about general chemistry, environmental fate, short- and long-term toxicity, ecological effects, and crop residues. In 1982, the EPA proposed additional changes to the data requirements, and in 1983, published technical guidelines for carrying out the required studies for registration. Registration review is based on an assessment of the potential effects of a product on human health and the environment when it is used according to the label. Because the labels are considered to have the force of law, use not in accordance with the label may result in civil or criminal penalties.

The 1988 amendments to FIFRA required that reregistration of products be completed more rapidly and imposed timelines for portions of the pesticide registration process. These amendments also made changes in EPA's funding, authorized fee collection for some new activities, and significantly altered responsibilities relating to suspended and canceled pesticides. These changes removed the requirement that EPA accept and dispose of canceled and suspended pesticides at EPA expense and eliminated certain clauses requiring EPA to provide compensation for the storage and disposal of suspended or canceled pesticides.

Other more recent amendments to FIFRA are discussed under the specific sections of the statute and will reference the associated codified regulations. The pesticide regulations are found in the *Code of Federal Regulations, Title 40, Protection of the Environment, Chapter 1, Environmental Protection Agency*, Parts 150 to 189 (40 CFR 150–189). In addition to the provisions already mentioned, FIFRA requires EPA and the States to establish programs to provide training to and certification of applicators, and to protect workers. It is important for surveillance program staff to have a basic understanding of FIFRA and the regulations that codify it. This will aid interagency cooperation since section numbers of FIFRA are often used in referring to types of enforcement inspections and pesticide product registrations. This section provides a brief summary of the sections of FIFRA pertinent to surveillance for pesticide-related illness.

SECTIONS 1 AND 2: DEFINITIONS

FIFRA Sections 1 and 2 are a table of contents and definitions. Selected definitions that are particularly important for surveillance are as follows:

"PEST—The term *pest* means (1) any insect, rodent, nematode, fungus, weed or (2) any other form of terrestrial or aquatic plant or animal life or virus, bacteria, or other micro-organism (except viruses, bacteria, or other micro-organisms on or in living man or other living animals) which the Administrator declares to be a pest under Section 25(c)(l)."

Note that the definition of *pest* is based only on humans deciding that a particular organism is deleterious or undesirable. An organism may be a pest to an individual or locale and may be considered a desirable part of the ecosystem in another location. This is particularly true for plant species where what is considered a weed in one area may be grown as a crop in another area.

"Pesticide—The term *pesticide* means (1) any substance or mixture of substances intended for preventing, destroying, repelling, or mitigating any pest and (2) any substance or mixture of substances intended for use as a plant regulator, defoliant, or desiccant, . . ."

Although the majority of the public tends to think of pesticides primarily as chemicals that kill pests, it is important to note that this definition in FIFRA also includes products that repel, prevent, and mitigate pests.

"To Use Any Registered Pesticide in a Manner Inconsistent with its Labeling—The term *to use any registered pesticide in a manner inconsistent with its labeling* means to use any registered pesticide in a manner not permitted by the labeling, except that the term shall not include (1) applying a pesticide at any dosage, concentration, or frequency less than that specified on the labeling unless the label specifically prohibits deviation from the specified dosage, concentration, or frequency, (2) applying a pesticide against any target pest not specified on the labeling if the application is to the crop, animal, or site specified on the labeling, unless the Administrator has required that the labeling specifically state that the pesticide may be used only for the pests specified on the labeling after the Administrator has determined that the use of the pesticide against other pests would cause an unreasonable adverse effect on the environment, (3) employing any method of application not prohibited by the labeling unless the labeling specifically states that the product may be applied only by the methods specified on the labeling, (4) mixing a pesticide or pesticides with a fertilizer when such mixture is not prohibited by the labeling, (5) any use of a pesticide in conformance with Section 5, 18, or 24 of this Act, or (6) any use of a pesticide in a manner that the Administrator determines to be consistent with the purposes of this Act. After March 31, 1979, the term shall not include the use of a pesticide for agricultural or forestry purposes at a dilution less than label dosage unless before or after that date the Administrator issues a regulation or advisory opinion consistent with the study provided for in Section 27(b) of the Federal Pesticide Act of 1978, which regulation or advisory opinion specifically requires the use of definite amounts of dilution."

This definition is a critical one for determining whether pesticides are used correctly, and it is frequently used at the State level when regulatory actions are taken.

SECTIONS 3, 4, AND 24(C): PESTICIDE REGISTRATION AND REREGISTRATION

The registration of pesticide products is covered under Sections 3 and 4 of FIFRA. The specific requirements for product registration are covered in 40 CFR Parts 152 to 167. As already mentioned this includes the review and "reregistration" of all pesticides. This also includes the review of pesticide product labels to ensure that they meet product labeling requirements codified in Part 156. Products must be registered if they meet the definition of a pesticide under 40 CFR, Section 152.15; and the product label or other materials indicate claims of pesticidal activity; or the product is represented in a manner that results in its being used as a pesticide. Requirements relating to worker protection are codified in 40 CFR 170; these will be discussed separately.

Products that are not considered pesticides and are exempt from FIFRA include those designed to control fungi, bacteria, viruses, microorganisms, or internal parasites/nematodes in living man or animals and are labeled for that purpose (e.g., pharmaceuticals). Other nonpesticides are plant nutrients, soil amendments, fertilizers, and disinfectants (including deodorizers and bleaching or cleaning agents) that do not make pesticidal claims.

Some additional products meet the definition of pesticides but are exempt from registration. These include pesticides regulated by another Federal agency, such as certain biological control agents and certain human drugs. Other exempted products with specific criteria are described in CFR 40 Section 152.25 and include treated articles or substances such as fabrics or paints where a registered pesticide is used to preserve the product itself; pheromones and pheromone traps meeting particular labeling criteria; preservatives for biological specimens and foods; natural cedar (not including oils, extracts, or mixtures); and minimum risk pesticides. Certain conditions allow pesticides to be transferred, sold, or distributed without registration as specified in CFR 40 Section 152.30.

Six registration types are provided for in FIFRA: new, amended, supplemental, reregistration, renewal, and Section 24(c) registrations. The requirements for submitting an application for registration are covered in CFR 40 Sections 152.40 through 152.175. The new registration is strictly that (that is, registration for a new pesticide product). An amended registration must be submitted if there is a proposed change in a product's composition, label, or packaging.

Reregistration and *renewal* refer to the processes for updating and reviewing previously approved uses on the basis of new data and standards. Section 4 of FIFRA covers the complex process established for reregistration.

Supplemental registrations allow distributors to assign their own brand name to a product that is registered by the producer. The product must be the same composition and be produced, labeled, and packaged in a registered establishment of the same producer. The supplemental product label must make the same claims as the primary product (some claims may be deleted), the product must be in the producer's container, and the registration number must have the distributor's company number added as a suffix.

Section 24(c) registrations are also referred to as *special local needs registrations*. These are issued when a State assigns an additional use for a federally registered product based on a local need. The State gives the product a special local need registration number and informs the EPA of this action within 10 days. The EPA then publishes the registration change in the Federal Register.

Section 5: Experimental Use Permits (EUPs)

EUPs are issued by the EPA to allow applicants to develop data needed to submit an application under FIFRA Section 3. These permits have specific terms, conditions, and time limitations and may be revoked at any time. A State with an approved State plan may issue EUPs.

Section 6: Administrative Review; Suspension

This section provides the EPA with the authority to cancel and/or suspend a pesticide registration. A registration must be canceled after 5 years if the registrant or an interested party does not request renewal of the registration before the end of the 5 years. The registrant is also required to submit additional information about adverse effects that is obtained after the registration (Section 6[a] [2]). The actual reporting requirements under this part of FIFRA were codified in 1997 by the addition of Part 159 to 40 CFR. The specific information that must be reported on toxic or adverse effect incident reports are addressed in Section 159.184. For more information about the 6(a)(2) regulations and reporting requirements, refer to http://www.epa.gov/pesticides/fifra6a2.htm.

The EPA may also issue a Notice of Intent to Cancel if there is information indicating that a pesticide, its label, or other required material does not comply with FIFRA. A hearing process and additional consultations are required as part of this process to cancel or reclassify a pesticide. Canceled pesticides that are already in the channels of trade may be sold or used after cancellation unless this is specifically prohibited in the cancellation order.

A suspension order may be issued after notifying the registrant and if EPA determines that there is an "imminent hazard" to human health posed by the continued use of a pesticide during the time necessary for cancellation. (Note: There may also be hearings as part of a suspension order.) Typically, the distribution, sale, or use is prohibited when a pesticide is suspended, although there are some circumstances when the sale or use of existing stocks is allowed.

SECTION 7: REGISTRATION OF ESTABLISHMENTS

This section requires that producers register establishments engaged in producing pesticide products or active ingredients used in these products. EPA assigns an establishment number that is associated with the name and address of the establishment. The types of records that must be maintained by establishments are described here. Specific annual reporting requirements exist for establishments and descriptions of what portion of required information is considered confidential.

SECTION 11: USE OF RESTRICTED USE PESTICIDES; APPLICATORS

The certification of applicators by EPA or State designees is described in this section. It includes requirements for making instruction on integrated pest management techniques available upon request. It establishes that there should be separate standards for commercial and private applicators.

SECTION 12: UNLAWFUL ACTS

A wide range of unlawful acts involving the sale, shipment, adulteration, registration, use, and testing of pesticides are described in this section. The FIFRA amendments of 1972 added the following language that designates the label as a legal document: "to use any registered pesticide in a manner inconsistent with its labeling."

SECTION 13: STOP SALE, USE, REMOVAL, AND SEIZURE

EPA or State designees are given the authority to issue stop sale, use, or removal orders if a pesticide or device is in violation of any portion of FIFRA. This includes pesticides that have been canceled by final order or suspended. The seizure of pesticides or devices through district court is prescribed for as described below:

- Pesticides that are adulterated or misbranded, not registered, improperly labeled, not colored, or discolored as required by FIFRA
- Product claims or directions for use are not those made in connection with its registration
- Misbranded devices
- A pesticide or device, when used in accordance with the requirements imposed by FIFRA and according to the label, still causes unreasonable adverse effects on the environment

Disposition of seized pesticides or devices, associated costs, and court costs are also covered in this section.

SECTION 18: EXEMPTION OF FEDERAL AND STATE AGENCIES

The EPA may exempt Federal or State agencies from any provision of FIFRA if it is determined that there are emergency conditions that require such an exemption. A Federal or State agency may request the determination from EPA that an emergency exists. EPA must consult with the U.S. Secretary of Agriculture and the governor of the concerned State when making the determination.

SECTIONS 22, 23, 26, AND 27: DELEGATION AND COOPERATION

Section 22 allows the Administrator to delegate authority to employees and directs the Administrator to cooperate with the U.S. Department of Agriculture (USDA) and appropriate State or political subdivisions to carry out the act and secure uniform regulations. Section 23 of FIFRA describes the authority for cooperative agreements between EPA and States. Under Sections 26 and 27, when information of an alleged violation of FIFRA or complaints are received by EPA, these will be referred to the State, and only "significant" complaints are tracked and followed by EPA. These sections authorize States to conduct inspections using Federal authority when such inspections are not authorized by State statutes. Funding of cooperative programs is shared by EPA and the State participants. The state may refer cases to EPA for Federal, civil, or criminal enforcement action.

WORKER PROTECTION STANDARD (WPS)

Under the 1972 revisions to FIFRA, changes required new wording on labels to increase worker safety, and in 1974, rules were promulgated that specifically addressed worker protection in 40 CFR 170. The 1974 changes were the addition of four basic requirements that pertained to hand laborers and included the following: workers should not be directly sprayed with pesticides; re-entry into treated areas was prohibited until dust was settled or spray dry with longer re-entry intervals for 12 specific pesticides; protective equipment was required for early re-entry; and "appropriate and timely warnings" regarding pesticide applications. The rules exempted many operations that use pesticides and did not include pesticide handlers. Reports of worker poisonings and concerns about the vagueness of the rules caused EPA to review these standards in 1983. An extensive process led to proposed revisions released in 1988. The final proposed rule was promulgated in 1992 and made changes to 40 CFR Part 170 and 40 CFR Part 156. These became fully effective in January of 1995.

Provisions of WPS include changes in labeling, an expanded scope of coverage to cover more workers and operations, prohibition of employer retaliation for attempting to comply with the standard, and requirements for

- warnings about pesticide applications
- use of personal protective equipment (PPE)
- restriction on re-entry into treated areas

- decontamination
- emergency assistance
- maintaining contact with handlers of highly toxic pesticides
- pesticide safety training

Following WPS enactment, EPA initiated a process to evaluate the impact of the standard. More information about WPS, compliance guides, training materials, and a *Field Inspection Pocket Guide* can be obtained from the EPA. (See Appendix G or go to http://www.epa.gov/pesticides/safety/workers/amendmnt.htm.) This Web site also contains information about EPA's ongoing assessment of WPS.

F.2 Federal Food, Drug, and Cosmetic Act (FFDCA)

This act requires the establishment of tolerances for the maximum amount of pesticide residues allowed in or on human food and animal feed products. It is jointly administered by EPA and the Food and Drug Administration (FDA).

F.3 Food Quality Protection Act (FQPA)

This law is a 1996 amendment to FIFRA and FFDCA, which changes the way standards are set for tolerances. It requires a health-based standard for assessing the risks of pesticide residues in food and feed. This standard differs significantly from past approaches in that it requires evaluating aggregated risk from dietary exposure as well as other nonoccupational sources of exposure such as drinking water and residential pesticide use. It also emphasizes the risks to infants and children. Additionally, EPA must consider cumulative exposure by evaluating the combined effects of different pesticides that may act in similar ways on the human body. The law also established a new standard for evaluating food-use pesticides: "reasonable certainty of no harm." This is in contrast to the previous standard of "no unreasonable adverse effects" that required a risk-benefit assessment. FQPA also requires reevaluation of all existing tolerances within 10 years. FQPA also

- includes authority to require testing for endocrine effects

- allows FDA to impose civil violations

- requires that a brochure on the health effects of pesticides be placed in grocery stores and allows States to require warnings or specific labeling of food that has been treated with pesticides

- prevents States from setting tolerances that are different from the Federal levels (*Note: There is a process for exceptions*)

- sets a 15-year cycle for pesticide reregistration renewal and provides additional funding for EPA processing of reregistration applications

- provides a faster registration process for new, safer pesticides

- establishes a minor use program to address EPA and USDA management of pesticides that have minor uses

- establishes new requirements for antimicrobial pesticides and clarifies jurisdiction
- establishes definitions and registration requirements for public health pesticides
- establishes definitions and regulation of two new categories of pesticide applicators: (1) maintenance applicators and (2) service technicians

Appendix G contains references for more information about FQPA.

F.4 Safe Drinking Water Act (SDWA)

SDWA was enacted to protect the quality of surface and below-ground sources of drinking water. It authorizes EPA to set standards for contaminants in drinking water including pesticides. As amended in 1996, it includes establishment for screening and testing programs for a variety of chemicals and pesticides, including endocrine effects of these chemicals.

F.5 Occupational Safety and Health Act (OSH Act)

The OSH Act was originally passed in 1970 and has undergone extensive amendments over the last 30 years. It provides for a safe and healthful working environment by giving authority to OSHA or its State partners to develop standards and enforce them. The act allows civil and criminal penalties to be issued for violations. There are 26 states with authorized State plans with standards that are at least as strict as Federal standards codified in 29 CFR Parts 1900–2400. State programs are based in a variety of State agencies depending upon the structure of State government; frequently the State OSH Act program is in the State bureau of labor or the State department of business and consumer affairs or equivalent. The State plans in Connecticut and New York cover only State and local government employees. Federal OSHA maintains offices in the non-State plan States to enforce provisions of the OSH Act.

At the Federal level, the interpretation of jurisdiction is that enforcement of issues associated with labeled pesticide products are within the purview of the EPA. State plans may have stricter interpretations and separate agreements with EPA or the State agency EPA designee responsible for enforcing FIFRA.

Responsibility for enforcing issues associated with the manufacture of pesticides prior to the placement of a FIFRA registered label on a container would fall under OSHA. The general industry-relevant portions of 29 CFR 1910 related to pesticide exposure are contained in

Subpart G—Occupational Health and Environmental Control (1910.94 to 1910.98)

Subpart H—Hazardous Materials (1910.101 to 1910.126)

Subpart I—Personal Protective Equipment (1910.132 to 1910.139)

Subpart J—General Environmental Controls (1910.141 to 1910.147 Appendix A)

Subpart K—Medical and First Aid (1910.151 to 1910.152)

Subpart Z—Toxic and Hazardous Substances (1910.1000 to 1910.1450 and Appendices), which includes air contaminants and hazard communication

A variety of other Federal regulations that involve pesticides are not highlighted in this manual since they may not be relevant to pesticide poisoning surveillance. These include, but are not limited to, the Resource Conservation and Recovery Act, the Clean Air Act, and the Endangered Species Act. Surveillance programs may need to become familiar with these acts or State laws related to them to address issues raised during particular case investigations. Other issues related to interagency coordination and overlapping jurisdictions are addressed in Chapters 5 and 6.

Appendix G
Resources for Additional Information Related to Pesticide Poisoning Surveillance

APPENDIX G ■

Resources for Additional Information Related to Pesticide Poisoning Surveillance

Contents

G.1   General Pesticide Resources
G.2   Recognition and Management of Pesticide Poisoning, Including Materials for HCPs
G.3   Pesticide Toxicology
G.4   Pesticide Products
G.5   Pesticide Usage
G.6   Pesticide Safety and Health Information to Assist Workers and Employers
G.7   Farmworker Employment, Demographics, Cultural Issues, and Service Organizations
G.8   Nonoccupational Exposure Issues (Homeowner, Schools, Vector Control, etc.)
G.9   State PPSP Contact Information
G.10 Federal Agency Contact Information
G.11 Agricultural Safety and Health (Other Than Pesticides)

This chapter serves as a starting point for any new PPSP to develop their own list of resources. Any compendium of this nature is always incomplete and becomes outdated. Efforts have been made to select references and resources that should remain available and are updated periodically. It includes resources that existing PPSPs find particularly useful. Listings and links are not an endorsement or guarantee of the accuracy of any Web site, publication, or other resource material. Also, any mention of specific products or laboratory resources are not an endorsement and are provided purely for informational purposes.

G.1 General Pesticide Resources

Several Internet Web sites are particularly useful when looking for information about any aspect of pesticides including regulation, toxicology, safety and health, and medical management. Links to most of the resources listed in this chapter can also be found at these sites.

California Department of Pesticide Regulation (CDPR)

http://www.cdpr.ca.gov/docs/dprdocs/docsmenu.htm

This site provides links to publications from the different CDPR programs. Subject areas include enforcement and worker safety (note that searches for publications including useful research reports may be done from a link in this section) and environmental monitoring.

CANADIAN CENTER FOR OCCUPATIONAL HEALTH AND SAFETY (CCOHS) HEALTH AND SAFETY INTERNET DIRECTORY

http://www.ccohs.ca/resources/hshome.html

This site has links to a large number of sites that address a broad range of occupational safety and health issues. It is divided into subject areas and also has a section on recently added links.

DUKE OCCUPATIONAL AND ENVIRONMENTAL MEDICINE

http://gilligan.duhs.duke.edu/oem/default.htm

This site provides many resources in the area of occupational and environmental medicine. A link connects to the National Library of Medicine to conduct literature searches. This page provides links to a moderated e-mail list that is used by many health care and public health professionals to exchange information and pose questions related to practice. Also at this site is a link to Duke's OEM WWW resource list that contains links to many of the sites referenced in this chapter.

PESTICIDE MANAGEMENT RESOURCE GUIDE

http://www.epa.gov/oppfead1/pmreg/

This EPA electronic resource compendium likely contains a link that can help you find what you need. The guide can be searched by subject, title, source, or full text. The general subject areas include alternative pest control methods (including IPM), bibliographic information, chemical and physical properties, chemical identification, environmental effects and fate, pesticide exposure in food, occupational pesticide exposure, formulation, information exchange, regulatory issue topics, training to develop pesticide regulatory staff, and worker safety. All listings contain summary information, how to access the resource, whether there is a cost, and some additional information about the resource when available (e.g., intended audience, language, updates, etc.).

UNIVERSITY OF NEBRASKA

http://pested.unl.edu/

This Web site contains a broad range of links related to pesticides. The links are categorized as follows: *IPM, Education, Health and Safety, Databases, Laws and Regulation, Newsletters, Environmental Protection, Organizations, and Product Manufacturers*. It also provides access to electronic versions of training materials from the Nebraska Agricultural Extension Service.

G.2 RECOGNITION AND MANAGEMENT OF PESTICIDE POISONING, INCLUDING MATERIALS FOR HCPs

This section provides information about resources related to the clinical recognition and management of pesticide poisoning. Resources Section G.3 *Pesticide Toxicology* also contain clinically relevant information.

G.2.1 PESTICIDES AND HUMAN HEALTH CONCERNS

G.2.1.1 TELEPHONE HOTLINES

LOCAL AND REGIONAL POISON CONTROL CENTERS (PCCs)

Local and regional PCCs provide case management information to HCPs who contact them. They provide assistance in identifying products and the constituent ingredients involved in poisonings. Poison centers also answer inquiries from the public, providing immediate first-aid information and referring individuals to seek medical consultation as needed. Poison centers also conduct public education campaigns to prevent poisonings. Some poison centers have additional functions including contractual relationships with certain manufacturers to provide information on product safety and health, and case management. In 2002, the American Association of Poison Control Centers (AAPCC) established a toll-free number (1-800-222-1222) that routes a caller to the PCC closest to them. Many PPSPs have found PCCs to be an important source of case reports. An up-to-date list of PCCs is available at http://www.aapcc.org/.

NATIONAL PESTICIDE INFORMATION CENTER (NPIC)

http://npic.orst.edu/index.html

NPIC is an EPA-funded resource available to the general public and HCPs. This is a toll-free telephone service that provides information about pesticide toxicology, environmental chemistry, and other product-specific information as well as recognition and management of pesticide poisoning. NPIC staff will refer callers to other resources as needed. NPIC's Web site provides ready access to EXTOXNET (the Agriculture Extension Service's library of information about pesticide toxicology) and to public consumer-oriented EPA publications. Telephone, Internet, electronic and mail access to fact sheets, brochures, and NPIC's annual reports are available from this resource.

Oregon State University
333 Weniger Hall; Corvallis, OR 97331-6502
E-mail: npic@ace.orst.edu
Telephone: 1-800-858-7378
Fax: 541-737-0761
Hours: 6:30 am-4:30 pm Pacific time, daily, except holidays.

The National Antimicrobial Network (NAIN) was the companion program addressing concerns about antimicrobial products (disinfectants, sterilants, and sanitizers), but it was discontinued March 31, 2002. For further information or assistance regarding antimicrobial products, contact the Antimicrobial Division in the EPA Office of Pesticide Programs by calling 703-308-0127, sending a fax to 703-308-6467, or e-mailing Info_Antimicrobial@epa.gov.

NATIONAL PESTICIDE MEDICAL MONITORING PROGRAM (NPMMP)

http://oregonstate.edu/npmmp/

NPMMP provides informational assistance in the assessment of human exposure to pesticides. NPMMP has a close relationship with NPIC. Both programs are based at the same location at OSU. Most of the inquiries received by NPMMP are referred from NPIC. When HCPs or others call NPIC in need of immediate medical information, they can usually be directly transferred from NPIC to the NPMMP. NPMMP also receives information requests from the general public and Federal and State agencies. NPMMP maintains a large paper and electronic library of information that is available upon request. Finally, NPMMP has limited funds to support the environmental investigation of suspected exposures using laboratory analyses.

G.2.2 PUBLICATIONS FOR HCPs

RECOGNITION AND MANAGEMENT OF PESTICIDE POISONING, 5TH ED. 1999

http://www.epa.gov/oppfead1/safety/healthcare/handbook/handbook.htm

JR Reigart, JR Roberts, eds. Publication No. EPA 735BRB98B003
Environmental Protection Agency; Office of Prevention, Pesticides and Toxic Substances
Ariel Rios Building, 1200 Pennsylvania Avenue, N.W., Washington, DC 20460
Cost: Free. (This publication is also available in Spanish.) To order 1 to 5 copies, call 703-305-7666 (Fax: 703-308-2962); for 6 or more copies, use the following:

U.S. EPA, National Service Center for Environmental Publications
PO Box 42419, Cincinnati, OH 45242-0419
Telephone: 800-490-9198
Fax: 513-489-8695
Internet: http://www.epa.gov/ncepihom

CASE STUDIES IN ENVIRONMENTAL MEDICINE

Department of Health and Human Services, Public Health Service, Agency for Toxic Substances and Disease Registry (ATSDR)

Self-instruction for primary HCPs. Continuing medical education (CME) credits are available through ATSDR. Pertinent case studies include the following:

- Arsenic
- Chlordane
- Cholinesterase inhibiting pesticides
- Pentachlorophenol
- Reproductive/developmental hazards

- Skin lesions
- Taking an exposure history (with an additional erratum document)

Cost: Free–There is a charge for CME. Some case studies can be downloaded in PDF format. An order form for case studies in print format can be downloaded from ATSDR at http://www.atsdr.cdc.gov/HEC/CSEM/status.html

To request copies of ATSDR Case Studies, call 404-498-0265, or contact:
Continuing Education Coordinator, ATSDR
Division of Health Education and Promotion, EB33
1600 Clifton Road, NE; Atlanta Georgia 30333

ENVIRONMENTAL HEALTH IN FAMILY MEDICINE

This is a set of self-teaching modules on aspects of environmental health pertinent to family medicine and primary care, which includes a pesticide training module with pediatric, adult, and geriatric cases. The curriculum can be downloaded from the International Joint Commission at http://www.ijc.org/rel/boards/hptf/modules/content.html.

GUIDELINES FOR PHYSICIANS WHO SUPERVISE WORKERS EXPOSED TO CHOLINESTERASE-INHIBITING PESTICIDES, 4TH ED., 2002

Pesticide and Environmental Toxicology Section, Office of Environmental Health Hazard Assessment (OEHHA), California Environmental Protection Agency, Oakland, CA.

Office of Environmental Health Hazard Assessment
Pesticide and Environmental Toxicology Section
1515 Clay Street, 16th Floor; Oakland, CA 94612
Telephone: 510-622-3170
Cost: Free–PDF version available at http://www.oehha.ca.gov/pesticides/programs/Helpdocs1.html

HANDBOOK OF PEDIATRIC ENVIRONMENTAL HEALTH, 2ND ED., 2003

RA Etzel, SJ Balk, eds.; American Academy of Pediatrics, Elk Grove Village, IL

This book is designed for pediatric HCPs and contains a chapter on pesticides.

American Academy of Pediatrics; PO Box 927,
141 Northwest Point Boulevard; Elk Grove Village, Il 60009-0927
Cost: $44.95. This book can be ordered online from the bookstore at http://www.aap.org/

OCCUPATIONAL SKIN DISEASE, 3RD ED., 1999

RM Adams, J Fletcher, eds. W B Saunders, Philadelphia, PA.

This is a useful reference that covers contact dermatitis, contact urticaria, systemic toxicity arising from percutaneous absorption, and other relevant topics.

APPENDIX G

PESTICIDE EXPOSURE AND THE HEALTH CARE PROFESSIONAL'S ACCESS TO PESTICIDE APPLICATION RECORD INFORMATION

http://www.ams.usda.gov/science/sdpr.htm

This brochure describes licensed HCPs' legal right of access to pesticide application records from certain pesticide applicators when the records may help in diagnosis and treatment of an exposed person. To obtain copies of this brochure or to request permission to customize the brochure with State-specific information, contact:

Pesticide Recordkeeping Program Staff
USDA Agricultural Marketing Service
Science and Technology, Pesticide Records Branch
87000 Centerville Road, Suite 202; Manassas, VA 20110
Telephone: 703-330-7826
E-mail: amspesticide.records@usda.gov

PESTICIDE POISONING SYMPTOMS AND FIRST AID, 2002

http://muextension.missouri.edu/xplor/agguides/agengin/g01915.htm
F Fishel, P Andre. MU Extension, University of Missouri-Columbia. Publication G 1915.
Electronic and PDF versions are available.
Cost: $.75-To order hard copies, call 1-800-292-0969.

PESTICIDE DERMATOSES, 2001

H Penagos, M O'Malley, H Maibach. CRC Press, Boca Raton, FL.

This is a comprehensive reference on skin reactions to pesticides, which includes diagnosis, treatment, and testing methods with case studies.

PESTICIDES AND EPIDEMIOLOGY, 2003

http://www.btny.purdue.edu/Pubs/PPP/PPP-43.pdf

F Whitford, J Acquavella, C. Burns. Purdue Pesticide Programs, Purdue University Cooperative Extension Service, West Lafayette, IN. Publication PPPB43.

A brief primer on pesticide epidemiology useful for explaining pesticide epidemiology to a wide variety of audiences.

G.2.3 Internet Resources

The following case study modules and clinical resources are available on the Internet.

Association of Occupational and Environmental Clinics (AOEC)

http://www.aoec.org/LLDIR.htm#Peer-reviewed%20Module

This association represents a network of clinics and occupational/environmental health specialists. Resources include a lending library of educational materials, environmental health-related course syllabi, and bibliographic references for HCPs. A helpful series of PowerPoint presentations on pesticide illness is available.

AOEC; 1010 Vermont Avenue NW, Suite 513
Washington, DC 20005
Telephone: 202-347-4976

Children's Environmental Health Case Study Series

http://www.uic.edu/sph/glakes/kids/case1/about.htm

This is from a project funded by AOEC, ATSDR, and EPA. A pediatric environmental health case study is included.

EnviroDx

http:/medlib.med.utah.edu/envirodx.

This is a multimedia computer-based learning program on environmental-related disease. It is a case-focused system described as a virtual clinic. The reader views video clips that provide the clinical history. Physical exam and test findings are provided. The reader is then provided an opportunity to select the diagnosis from a multiple choice list. A video clip of the prevention counseling provided to the patient is also available.

EnvirRn

http://envirn.umaryland.edu/

This Web site offers a broad spectrum of information about environmental health from a nursing perspective and includes educational and resource links. A pesticide case study is on this Web site (http://envirn.umaryland.edu/interventions/pestcase.htm).

Pesticide Poisoning Diagnostic Tool (Poisoning Database)

http://www.pesticideinfo.org/Search_Poisoning.jsp

This tool, developed by PANNA, is designed to help HCPs and others recognize, diagnose, and report pesticide-related illnesses. The database currently provides symptom, first aid, and treatment-related information for approximately 1,900 pesticides. It can be searched for possible pesticide poisoning agents by entering as little or as much relevant information as is available. Searches by chemical or product name, pesticide use type, geographic location, and/or crop or application site are possible. In addition, a HCP (or other user) can search by observed signs and symptoms. This online resource also provides reporting information (legal requirements, reporting instructions, and official reporting contacts) for all 50 States. The database lists specific county-level reporting information for California and Florida.

Pesticide Use on Airlines

http://ostpxweb.dot.gov/policy/Safety%20Energy%20Env/disinsection.htm

There is concern about pesticide exposures of workers and passengers in the air transportation industry. Some countries require arriving international flights to undergo disinsection. Disinsection involves either spraying aerosol insecticide in the cabin with airline passengers present, or applying a residual insecticide to the cabin when passengers are not present. The Web site provides more information on this topic, including a list of airline contacts and a list of countries that require disinsection.

Specific Medical Tests Published in the Literature for OSHA Regulated Substances

http://www.cdc.gov/niosh/nmed/medstart.html

This resource on medical tests for particular chemicals is on the NIOSH Web site. This site provides information about specific medical tests that have been published for OSHA regulated substances, and includes some pesticides.

G.2.4 Additional Resources Related to HCP Training on Recognition and Management of Pesticide Poisoning

This section includes background material related to curricula for health provider training programs.

NIOSH Agricultural Health and Safety Centers

The NIOSH Agricultural Health and Safety Centers offer seminars (usually jointly sponsored with other partners) on pesticide medicine. They can provide information about the curriculum used for those seminars. See Section G.11 *Agriculture Safety and Health* for contact information.

PESTICIDES AND NATIONAL STRATEGIES FOR HCPS

http://www.neetf.org/Health/providers/index.shtm

There is an ongoing national initiative for increasing HCP education and training to improve recognition, management, and prevention of pesticide-related health conditions. This is a long-term, collaborative process created by the EPA, U.S. Department of Health and Human Services (DHHS), U.S. Department of Agriculture, and the National Environmental Education and Training Foundation (NEETF). The initiative advocates the education of health professionals while they are in school as well as developing model programs and practice guidelines to incorporate pesticide poisoning recognition, management, and prevention into the health care setting. The development and dissemination of training curricula and other resources targeted at practicing HCPs have also been proposed. To obtain *Pesticides and National Strategies for HCPs: Pesticides Initiative, National Pesticide Competency Guidelines for Medical & Nursing Education,* or *National Pesticide Practice Skills Guidelines for Medical and Nursing Practice* and other publications as they become available, contact NEETF.

National Environmental Education & Training Foundation
1707 H Street, NW; Suite 900; Washington, DC 20006
Telephone: 202-261-6481
http://www.neetf.org

The NEETF Pesticides Resource Library Web site is

http://www.neetf.org/Health/Pestlibrary.shtm

G.2.5 PESTICIDES AND ANIMAL HEALTH CONCERNS

NATIONAL ANIMAL POISON CONTROL CENTER

http://www.aspca.org/site/PageServer?pagename=apcc

This is a nonprofit PCC that addresses concerns and provides veterinary diagnostic and treatment information for poisoned animals. It requires a fee-for-service, which is paid by the animal owner, veterinarian, or a product manufacturer.

ASPCA National Animal Poison Control Center (member of AAPCC)
1717 South Philo Road; Suite 36 Urbana, IL 61802
Emergency Telephone: 888-426-4435

G.3 PESTICIDE TOXICOLOGY (ALSO SEE SECTION G.4 PESTICIDE PRODUCTS)

This section lists toxicological resources that provide information about pesticides. The listed resources include a few books that the PPSP may choose to include in a basic library. The Internet-based data resources can be accessed free of charge. Many other proprietary Internet-based data systems are not listed here.

G.3.1 PUBLICATIONS

CASARET AND DOULL'S TOXICOLOGY: THE BASIC SCIENCE OF POISONS, 6TH ED., 2001

CD Klaassen, ed. McGraw-Hill, Medical Publishing Division, New York, NY.

This is a standard toxicology text that includes chapters on pesticide toxicology and occupational toxicology, in addition to sections on principles of toxicology and other aspects of toxicology.

CLINICAL ENVIRONMENTAL HEALTH AND TOXIC EXPOSURES, 2ND ED., 2001

JB Sullivan, GR Krieger, eds. Lippincott Williams & Wilkins, Philadelphia, PA.

This clinical reference on environmental toxicology has sections on workplace and emergency responses to hazardous material exposures and information about specific exposures including pesticides.

HANDBOOK OF PESTICIDE TOXICOLOGY, 2ND ED., 2001

RI Krieger, ed. Academic Press, San Diego, CA

This is a two-volume set that covers a wide variety of topics and includes toxicology reviews on many classes of pesticides.

THE PEST BOOK, 5TH ED., 2000

GW Ware. Thomson Publications, Fresno, CA

This reference covers chemistry, mode of action, and issues related to handling of pesticides.

TOXICOLOGY OF THE EYE: EFFECTS ON THE EYES AND VISUAL SYSTEM FROM CHEMICALS, DRUGS, METALS AND MINERALS, PLANTS, TOXINS, AND VENOMS, 4TH ED., 1999

WM Grant, JS Schuman. Charles C. Thomas Pub., Ltd., Springfield, IL.

This reference on toxicology of the eye includes information about some pesticides and some chemicals used as inerts and carriers in pesticide products.

G.3.2 INTERNET DATA RESOURCES

The sites listed here contain toxicological data on pesticides. While many other sources are available, those listed provide links to other sources and are a good starting point.

EPA REREGISTRATION ELIGIBILITY DECISION DOCUMENTS (REDs)

The REDS can be a useful source of toxicology information since they contain regulatory reviews of pesticides that were first registered before November 1, 1984. FIFRA requires that these active ingredients be reviewed to determine if they can be reregistered (that is, they must not cause unreasonable risks to people or the environment when used in accordance with the approved label). These documents can be accessed from the EPA Office of Pesticide Programs home page at http://www.epa.gov/pesticides, or be ordered from the National Service Center for Environmental Publications (NSCEP).

EXTOXNET

http://ace.ace.orst.edu/info/extoxnet/

EXTOXNET–The EXtension TOXicology NETwork–is a useful source of toxicology-related information about pesticides. The toxicological information in this data system is developed cooperatively by the University of California-Davis, Oregon State University, Michigan State University, Cornell University, and the University of Idaho.

Hazardous Substances Data Bank (HSDB)

http://toxnet.nlm.nih.gov

The HSDB is one of the toxicology data files in the National Library of Medicine's Toxicology Data Network (TOXNET7). It includes information about human exposure, industrial hygiene, emergency handling procedures, environmental fate, and regulatory requirements. The sections on treatment include much of the information that is found in the POSINDEX7 data system used by PCCs.

INCHEM

http://www.inchem.org/

This site is produced by the International Programme on Chemical Safety (IPCS), which is a cooperative program of the World Health Organization (WHO), International Labor Organization (ILO), and the United Nations Environment Programme (UNEP). It provides access to several series of documents produced by WHO and its partners. The information sources that will likely be of greatest use to PPSPs are described below. Additionally, this Web site has International Agency for Research on Cancer (IARC) summaries, evaluations of the toxicity of pesticide residues in foods, and information on non-pesticide chemicals and pesticidal inert ingredients.

Environmental Health Criteria Monographs

This is a series of monograph publications from the WHO. Many of these monographs describe the characteristics and effects of pesticides on humans and animals. Each monograph on a class of compounds covers physical properties and analytical methods, sources of exposure, information about environmental transport and fate, review of human and animal health effects, and risks.

Health and Safety Guides

A more simply worded general use series on a range of chemicals that includes risks of exposure, summary information about effects, and medical and administrative issues related to exposure.

PESTICIDE DATA SHEETS

These are peer-reviewed fact sheets that give basic toxicologic information about pesticides in broad worldwide use, as well as those that have been found to be particularly hazardous. There is some duplication with fact sheets available from the EXTOXNET site. It also includes information about products registered for use outside of the United States.

POISON INFORMATION MONOGRAPHS

This peer-reviewed series emphasizes the health effects related to exposure from a variety of pesticides and other toxins. It includes information about patient evaluation and management.

INTEGRATED RISK INFORMATION SYSTEM (IRIS)

http://www.epa.gov/iriswebp/iris/index.html or http://toxnet.nlm.nih.gov

IRIS is a database on the human health effects that may arise from environmental chemical exposure. It provides specific hazard and dose response information. It is designed for users with knowledge about toxicology and life sciences. This system also provides information to aid users in accessing and understanding IRIS data.

PAN PESTICIDE DATABASE

http://www.pesticideinfo.org/Index.html

This database has been compiled by PANNA and includes peer-reviewed scientific information about pesticide products and ingredients. Information about acute and chronic health effects, and ecotoxicity is summarized from a number of sources including the EPA, IRIS, National Toxicology Program, International WHO, National Toxicology Program (NTP), National Institutes of Health (NIH), International Agency for Research on Cancer (IARC), the European Union (EU), and the State of California.

G.4 PESTICIDE PRODUCTS

G.4.1 Databases on Pesticide Products

These data systems provide the PPSP with the ability to search and identify pesticide products, formulations, ingredients, and regulatory status.

THE CALIFORNIA PESTICIDE DATABASES—USEPA/OPP PESTICIDE-RELATED DATABASE QUERIES

http://www.cdpr.ca.gov/docs/epa/epamenu.htm

The California Department of Pesticide Regulation (CDPR) has developed a query system using data from the US EPA Office of Pesticide Programs (OPP) pesticide product information system (PPIS). (The PPIS system is also used to develop the pesticide ingredient and product information used in the

SPIDER surveillance database system.) The databases found at this site include a pesticide product database, a chemical ingredient database, and a pesticide manufacturer database. In addition, links are provided to pesticide label images located at EPA.

PAN Pesticide Database

http://www.pesticideinfo.org/Index.html

This database has been compiled by the PANNA and includes peer-reviewed scientific information about pesticide products and ingredients. The system can be searched by pesticide product name or registration number. A variety of advanced searches can be performed (e.g., a search for all products containing a particular active ingredient). Information about acute, chronic and ecotoxicity is summarized from a number of sources including the EPA, IRIS, National Toxicology Program, WHO, National Toxicology Program (NTP), National Institutes of Health (NIH), International Agency for Research on Cancer (IARC), the European Union (EU), and the State of California.

G.4.2 Pesticide Product Labels

There are several Internet sources for obtaining pesticide product labels. In addition to these sources, the label as registered in each State is available from the State department of agriculture (or other designee responsible for registration).

EPA pesticide product label system

http://www.epa.gov/pesticides/pestlabels/

The EPA pesticide product label system is a collection of images of product labels that have been approved by OPP. (Labels are in TIFF format.) Apart from the label, additional correspondence and amended labels are included in the system. Because the system is indexed by the product registration number, you will first need to enter this number. The CDPR site is a useful source of registration numbers, and it can be searched for all federally registered products (and their corresponding registration number) by active ingredient, product name or company name (http://www.cdpr.ca.gov/docs/epa/epamenu.htm).

Greenbook

http://www.greenbook.net

The *Greenbook* is produced by the Chemical and Pharmaceutical Press. Labels, supplemental labels, and MSDSs are obtained from the pesticide companies and compiled into a variety of formats (book, CD-ROM, Internet access).

G.4.3 Material Safety Data Sheet (MSDS) Directories

In addition to the *Greenbook* site described above, several other sites contain pesticide product MSDSs. Some entries in Section G.1 also contain links to MSDS directories. PPSPs have found the following sites to be useful:

CDMS Label / MSDS Information

http://www.cdms.net/manuf/manuf.asp

Vermont SIRI MSDS Collection

http://www.siri.org/msds/links.html

Where to find MSDS on the Internet

http://www.ilpi.com/msds/index.html

There is no directory of Spanish language MSDSs for pesticide products registered in the United States. Some manufacturers have MSDSs in Spanish for particular products that are available on request. These MSDSs are not posted on Web sites, and therefore require a telephone or written request to the manufacturer. See the following link for further information about this issue: http://www.ilpi.com/msds/faq/parte.html#foreign.

G.4.4 Pesticide Manufacturers

The following sites contain links to pesticide manufacturers:

CropLife America

http://www.croplifeamerica.org

This is a nonprofit trade organization for the crop protection, pest control, and biotechnology industries. Their Web site has links to many pesticide manufacturers.

NPIC

http://npic.orst.edu/manuf.htm

PestWeb

http://www.pestweb.com/

This is a pest control industry Web site that contains links to many manufacturers.

University of Nebraska

http://pested.unl.edu/

G.5 Pesticide Usage

Information about pesticide use can be helpful during the process of developing a PPSP and in analyzing and evaluating PPSP data. There is no comprehensive source of pesticide use information. Some systems collect data on a segment of pesticide applications while others conduct periodic surveys to obtain data and create estimates of pesticide usage. Six States mandate some form of pesticide use reporting: Arizona, California, New Hampshire, New York, New Jersey, and Oregon. Not all of these State systems provide data in a readily accessible format. Other States are in varying stages of developing or exploring the development of pesticide use reporting systems. This section will briefly describe some sources of available data and reviews of data. Other sources of data not described here include Federal and State systems that monitor pesticide residues in food.

G.5.1 Data Sources

California Environmental Protection Agency, CDPR

http://www.cdpr.ca.gov/docs/dprdocs/docsmenu.htm

CDPR within the California Environmental Protection Agency has maintained a database since 1970, which includes use of restricted use pesticides by farmers and all applications by pest control operators. In 1990, the system changed to require that all agricultural pesticide use be reported on a monthly basis. Agricultural use is broadly defined within this system and includes right-of-way, park, and golf course applications. An overview of this pesticide use reporting system and reports are under the topic heading of Pesticide Sales and Use. This data is also accessible at the PANNA site (http://www.pesticideinfo.org/Search_Use.jsp).

EPA

http://www.epa.gov/oppbead1/pestsales/

EPA has conducted surveys to obtain information about homeowner and commercial pesticide use. This data and USDA agricultural pesticide-use data have been combined to produce estimates of pesticide use in agricultural and nonagricultural sectors. The most recent sales and usage reports are available from the NSCEP or on the web as a PDF file.

New York Department of Environmental Conservation (NYDEC)

http://www.dec.state.ny.us/website/dshm/pesticid/prl.htm

This State enacted pesticide reporting legislation in 1996. The NYDEC initiated data collection February 1, 1997. The system collects information about pesticide use by commercial pesticide applicators, sales of restricted-use pesticides to farmers, and sales of restricted-use pesticides to commercial applicators and dealers. A description of the program and a link to data reports is available at their Web site.

UNITED STATES GEOLOGIC SURVEY, NATIONAL WATER QUALITY ASSESSMENT PROGRAM

http://water.usgs.gov/nawqa/

This program started in 1991 and assesses trends in water quality. Pesticides are considered an important issue in this program. Data and maps describing pesticides in streams, groundwater, and sediments, along with methodology, are available.

USDA

National Agricultural Statistics Service (NASS)

http://www.usda.gov/nass/offices.htm

The NASS conducts a variety of surveys on the use of pesticides and fertilizers on farms and ranches. Annual reports are available from this system that list chemical application rates and acres treated for major crops. Information about some fruit and vegetable crops is reported in alternate years. The NASS has 45 local field offices that work with State agencies and organizations to collect survey data. State reports are released on a periodic basis. A listing of State NASS offices and access to reports is available at the NASS Web site.

CENSUS OF AGRICULTURE

http://www.nass.usda.gov/census

Originally conducted by the Census Bureau, the Census of Agriculture collects data on agricultural production and provides data down to the county level. It has been conducted in some form since 1840, initially every 10 years. It has been conducted every 5 years since 1925, currently on a schedule of years ending in 2 and 7. The survey collects information about land use, ownership, acreage, production, economic information, number of hired farmworkers, injuries and deaths, and chemical, fertilizer and machinery use. Available reports include an atlas of maps prepared from these data.

OTHER USDA REPORTS

http://usda.mannlib.cornell.edu/ess_entry.html

Other relevant agricultural reports can be obtained from the USDA Economics and Statistics System, which incorporates a variety of USDA datasets.

G.5.2 REVIEWS OF PESTICIDE USE DATA SOURCES

Two documents have reviewed the availability of pesticide use data in the last few years. These reviews were done for different purposes. Each describes available data and gaps in data.

Appendix G

Oregon Pesticide Use Reporting System: Analytical Review. May 2000

The information presented in this report was gathered and formatted by staff of the Oregon Health Sciences University (OHSU) and Oregon State University (OSU), under contract with the Oregon Department of Agriculture (ODA). The report provides information to be considered by ODA in the development and implementation of the Oregon Pesticide Use Reporting System and can be obtained from ODA or downloaded from http://oda.state.or.us/purs/history/anreview/index.html.

Oregon Department of Agriculture, Pesticides Division
635 Capitol Street N.E.; Salem, OR 97301-2532
Telephone: 503-986-4635

The Toxic Treadmill. Pesticide Use and Sales in New York State, 1997–1998

Audrey Thier, Environmental Advocates, and New York Public Interest Research Group. October, 2000 (Revised 3/29/01). This report reviews and critiques pesticide use reporting in New York State and is available at http://www.eany.org/.

G.6 Pesticide Safety and Health Information to Assist Workers and Employers

A large body of information is available on this subject. This section highlights some of the core publications, electronic documents, and electronic listings of training materials and other resources.

American Association of Pesticide Safety Educators

http://aapse.ext.vt.edu/aapse.html

This is an association of persons and organizations involved in providing education on pesticide safety. The Web site provides links to State resources, a speakers bureau, EPA regulations, newsletters, and journals. It has tools for developing and presenting training programs and also provides access to materials for evaluating pesticide applicator training programs.

Association of Farmworker Opportunity Programs (AFOP)

http://www.afop.org

Radio Pesticida. Set of six cassettes with a guide, includes five mini-dramas and five accompanying talk shows addressing drift, exposure, and safety at home and work.

Cost: $50.00. Tapes are in Spanish, and the training manual is in English and Spanish.
    Radio Pesticide, Haitian Creole version: $25.00
Telephone: 703-528-4141

BIBLIOGRAPHY FOR TRAINERS OF FARMWORKERS UNDER THE WORKER PROTECTION STANDARD, 2000

Melissa Frisk, Institute for Agriculture and Trade Policy.

This document is an annotated listing of more than 75 resources for training farmworkers in pesticide safety. Contact information for ordering resources is included. (45 pages)

Cost: $3.00 (including postage). Available online as a PDF document.

Contact: Candace Falk, Institute for Agriculture and Trade Policy (IATP)
2105 First Avenue South; Minneapolis, MN 55404
Telephone: 612-870-3453
Fax: 612-870-4846
E-mail: cfalk@iatp.org
Web site: http://www.iatp.org/labels/ (Click on the *Resources* button on the left, then click on the *library* tab near the top and select the pull-down selector for *by IATP staff* to locate the document.)

CALIFORNIA DEPARTMENT OF PESTICIDE REGULATION (CDPR)

The CDPR Worker Health and Safety Branch produce the *Pesticide Safety Information* series to assist employers to comply with regulatory training requirements. The series is available in English and Spanish and covers both crop and noncrop settings. All of the documents are available in PDF format from http://www.cdpr.ca.gov/docs/whs/psisenglish.

The *Worker Health and Safety Reports* are an additional series of publications that address other worker health and safety issues. A search for these reports can be made using a keyword or publication number at http://www.cdpr.ca.gov/docs/whs/whsrep.htm.

CROPLIFE AMERICA

www.croplifeamerica.org

A crop protection, pest control, and biotechnology industry nonprofit trade organization site that contains worker safety and health information under its *Stewardship* and *Publication* links. Topics include safe work practice information in English and Spanish and storage and disposal information. Guidance documents designed as tips for pesticide users are also available. A list is available of things a grower should be aware of when working with aerial applicators, including the need to provide advance notification to neighbors and workers.

EPA

http://www.epa.gov/pesticides/health/worker.htm

The EPA has many educational resources available on worker safety and training.

PESTICIDE EDUCATION–UNIVERSITY OF NEBRASKA

http://pested.unl.edu/

This Web site provides links to the Nebraska Agricultural Extension pesticide safety education resources as well as links to other sites.

PESTICIDE SAFETY TEACHING RESOURCES, VA TECH LIST

http://www.vtpp.ext.vt.edu/htmldocs/trainres.html

This Web site contains links to resources for teaching pesticide safety.

THOMSON'S SPANISH-ENGLISH ENGLISH-SPANISH, ILLUSTRATED AGRICULTURAL DICTIONARY

RP Rice, Jr. [1993] Thomson Publications, Fresno, CA

An illustrated dictionary of terms used in agriculture, including tools, irrigation, animals, crops, and plant propagation, with a broader non-illustrated section that includes terms related to pesticides and pesticide application.

UNIVERSITY OF CALIFORNIA STATEWIDE IPM PROGRAM

http://anrcatalog.ucdavis.edu

This program offers a wide range of high quality publications on integrated pest management (IPM) and safe application of pesticides. A few publications of interest are listed here by title, with year of publication. The relevant publications are listed under the headings of *Pest and Disease Management* and *Pest Control Training and DPR Test Materials* in the Agriculture and Natural Resource (ANR) Catalog.

- *Establishing IPM Policies and Programs: A Guide for Public Agencies, 2003.*

- *La seguridad en el manejo de pesticidas, 1999.*

- *Illustrated Guide to Pesticide Safety, 1999.* (Instructor version; worker versions in English, Spanish, and Punjabi)

- *Jorge's New Job: Getting Tested for Cholinesterase, 2000.* (English and Spanish)

- *La lotería de los pesticidas, 1992.* (Spanish/English pesticide safety training game with trainer's manual)

University of California; Agriculture and Natural Resources (ANR) Communications Services
6701 San Pablo Avenue; Oakland CA 94608-1239
Telephone: 1-800-994-8849
Fax: 510-643-5470

G.7 Farmworker Employment, Demographics, Cultural Issues, and Service Organizations

G.7.1 Employment Issues

Migrant and Seasonal Agricultural Worker Protection Act (MSPA) (29USC[1], et seq.)

This act provides legal protections for migrant and seasonal agricultural workers in their interactions with labor contractors, employers, and providers of housing. The act addresses issues of labor contract registration and disclosure of terms of employment and housing occupancy. A poster lists the rights and protections of this group of workers. The Wage and Hour Division of the United States Department of Labor (DOL), along with its State level designees, is responsible for enforcing the MSPA. More information about the MSPA and the poster are available at
http://www.dol.gov/esa/whd/mspa/index.htm.

Workers' Compensation Coverage for Agricultural Workers

State laws vary widely on whether agricultural workers are covered by workers' compensation. The two reports listed here provide some background information about this issue.

Legal Background Paper: Protection of Migrant Agricultural Workers in Canada, Mexico, and the United States. 2002
http://www.naalc.org/english/study4.shtml
The Secretariat of the Commission for Labor Cooperation. Washington DC.

Increasing Farmworkers' Access to Workers Compensation Benefits.
S Davis, Farmworker Justice Fund, Inc., 1999. PC499
Report available from HRSA Contact: 1-800-BASKBHRSA or http://www.ask.hrsa.gov/

G.7.2 Child Labor in Agriculture

PPSP staff should be familiar with the agricultural exemptions of the Fair Labor Standards Act and other factors influencing the safety and health of children working in agriculture. The following publications provide a starting point for becoming familiar with the issue.

Child Labor in Agriculture: Changes Needed to Better Protect Health and Educational Opportunities, 1998. (GAO/HEHSB98B193.) U.S. General Accounting Office, Washington, DC; http://www.gao.gov

Fingers to the Bone: United States Failure to Protect Child Farmworkers, 2000. Human Rights Watch, Washington, DC. Telephone: 202-612-4321; http://www.hrw.org/

The Ones the Law Forgot: Children Working in Agriculture. Shelley Davis and James B. Leonard, Farmworker Justice Fund; http://www.fwjustice.org/fjf_reports.htm.

Pesticides: Improvements Needed to Ensure the Safety of Farmworkers and Their Children, 2000. (GAO/RCEDB00B40.) U.S. General Accounting Office, Washington, DC; http://www.gao.gov.

G.7.3 FARMWORKER DEMOGRAPHICS

Changes in agricultural work, immigration patterns, and government policies all impact the demographics of agricultural workers in the United States. These demographics are important to consider when developing programs, interventions, and studies. A few sources of information are provided here.

CALIFORNIA INSTITUTE FOR RURAL STUDIES (CIRS)

Suffering in Silence: A Report on the Health of California's Agricultural Workers. Nov 2000.

In 1999, the CIRS conducted the study described in this report. The report provides baseline data on the health status of California's agricultural workers. Included is a health status assessment based on a medical exam with blood chemistry, information about access to health care, occupational injuries, workers' compensation coverage, pesticide safety training, and sanitation. The report can be obtained from CIRS or downloaded from the Internet. Further analysis of these data and follow-up studies will be available from CIRS.

California Institute for Rural Studies
P.O. Box 2143, Davis, CA 95616
Telephone: 530-756-6555
Web site: www.cirsinc.org

ENUMERATION STUDIES

The Migrant Health Program at the Bureau of Primary Care, Health Resources and Services Administration, U.S. Department of Health and Human Services (DHHS), has contracted for enumeration studies of migrant seasonal farmworker populations at the county level in ten States (Arkansas, California, Florida, Louisiana, Maryland, Mississippi, North Carolina, Oklahoma, Texas, and Washington). Profiles may be developed for additional States. The Migrant Health Program definition of migrant and seasonal farmworker was used for this study. The scope does not include all sectors of the agricultural industry or year-round nonmigrant agricultural workers. The reports are available from their office (see Section G.10 for the address and telephone number), at their Web site (http://bphc.hrsa.gov/migrant/), or at the Web site of the National Center for Farmworker Health (http://www.ncfh.org).

NATIONAL AGRICULTURAL WORKER SURVEY (NAWS)

This extensive survey is conducted by the US DOL. It collects information on farmworkers, including information about household members, detailed demographics on the farmworker, detailed employment information (e.g., location and work type), wages, working conditions, availability of safety and health training and equipment, household income, legal status, and use of social services. The questionnaire and periodic reports can be obtained from the DOL at http://www.dol.gov/asp/programs/agworker/naws.htm or by telephone at 202-219-6197.

G.7.4 Farmworker Cultural Issues

The three resources listed here explore some pertinent cultural issues. Farmworkers come from culturally diverse backgrounds. By understanding the farmworker's perception of risk, safety behavior, and use of health care, valuable insight can be gained on the effect of these beliefs on the surveillance of pesticide-related illness and injury.

BIBLIOGRAPHY OF BOOKS ON MIGRANT WORKERS

http://www.fwjustice.org/bibliography.htm

The Farmworkers Justice Fund has developed a selected bibliography of books, both fiction and nonfiction, on migrant agricultural workers. The listing includes current and older works that provide insight into the historical, social, and cultural aspects of migrant agricultural work.

ETHNOMED

http://www.ethnomed.org

This Web site provides information about cultural beliefs and medical issues that relate to health care. The cultural groups included are recent immigrant groups in the Seattle area since the site is located at the Harborview Medical Center, University of Washington. PPSP staff may find useful information in the Cultural Profiles and at some of the related Web site links.

MIGRANT CLINICIANS NETWORK (MCN)

http://www.migrantclinician.org/excellence/cultural/

Dr. Jennie McLuarin has written two pieces on cultural competency for *Streamline*, the MCN newsletter (March/April 2002 and November/December 2002). The articles and other cultural competency resources can be accessed from the Web address given above.

G.7.5. FARMWORKER SERVICE ORGANIZATIONS

ASSOCIATION OF FARMWORKER OPPORTUNITY PROGRAMS (AFOP)

http://www.afop.org

AFOP is an alliance of organizations serving farmworkers and their families. The services provided by AFOP include information, education, support, advocacy, and representation at the national level.

Since 1995, AFOP has had a collaborative program with AmeriCorps to support work in rural farmworker service agencies. The US EPA provides these farmworker service agencies with curriculum materials to promote pesticide safety. AFOP's national headquarters coordinates the program activities and provides expert technical assistance and training. By the end of 2000, the program had trained more than 215,000 farmworkers, provided at least 82,000 community services, and placed some 450 AmeriCorps members in communities throughout the United States. For more information about this program, contact 703-528-4141. The *ASAFE: Serving America's Farmworkers Everywhere* link on the AFOP Web page provides information about AFOP members who participate in the program in Arizona, Arkansas, California, Florida, Maine, Maryland, Massachusetts, New Jersey, New York, North Carolina, Ohio, Oregon, Pennsylvania, Tennessee, Virginia, and Washington.

FARMWORKER JUSTICE FUND (FJF)

http://www.fwjustice.org/

FJF is a nonprofit organization that works to improve working and living conditions for migrant and seasonal farmworkers in the United States. FJF provides training, legal advocacy, and technical assistance. Their Web site provides links to farmworker labor organizations, migrant farmworker labor law sites, and GAO reports on farmworkers.

LEGAL SERVICES CORPORATION (LSC)

http://www.lsc.gov/index2.htm

LSC is a private nonprofit corporation established by Congress in 1974 to provide civil legal assistance to those who are otherwise unable to afford it. A listing of LSC-funded programs and links to programs with Web sites is available on the homepage listed above. The local legal services offices were mentioned in Chapter 3 as a source of reports, since farmworkers with concerns about pesticide exposures and workplace safety may seek legal assistance.

MEXICAN CONSULATES

http://www.mexonline.com/consulate.htm

The consulates can be valuable partners for outreach and education activities aimed at Mexican national and immigrant agricultural workers.

NATIONAL CENTER FOR FARMWORKER HEALTH (NCFH)

http://www.ncfh.org

This is a nonprofit organization that serves farmworkers and farmworker health service agencies and organizations. It sponsors four annual conferences that focus on the health issues impacting the different migratory streams traveled by farmworkers in the United States. This organization has a catalogue of products, many of which are free or low cost. Materials include videos, resources for HCPs, patient education materials, a bibliography, and some research information. This group also sponsors an electronic discussion group on migrant health research issues. This listserv encourages the exchange of information, resources, grant opportunities, and original research in the area of migrant health. Subscriptions to the listserv are available on the organization's homepage.

G.8 NONOCCUPATIONAL EXPOSURE ISSUES (HOMEOWNER, SCHOOLS, VECTOR CONTROL, ETC.)

G.8.1 PUBLIC CONSUMER INFORMATION

A broad body of information is available for the general public on the safe use of pesticides. Most of the sources have already been mentioned in other sections of this appendix. All of the sites listed in the General Resources section contain some materials aimed at the general public. NPIC's Web site provides ready access to EXTOXNET, the Agriculture Extension Service's library of information about pesticide toxicology and to consumer-oriented EPA publications (see Section G.3.2). The EPA-OPP Web site home page has additional consumer resources including fact sheets on drift and pesticides in the home (www.epa.gov/pesticides). Some topic-specific areas of information for the public are listed below.

G.8.2 PESTICIDES IN SCHOOLS

The use of pesticides in and around schools has been an area of public concern in the last few years. Several States have passed legislation addressing the use of pesticides in schools, including the adoption of parental notification. Links to some recent reports on pesticide use in schools and resources on IPM in schools are provided below:

CALIFORNIA SCHOOL INTEGRATED PEST MANAGEMENT PROGRAM

http://www.cdpr.ca.gov/cfdocs/apps/schoolipm/main.cfm

This site also has several resources on school IPM. Among them is this report:

Overview of Pest Management Policies, Programs, and Practices in Selected California Public School Districts. March 1996. PM 96-01.
Sewell E. Simmons, Timothy E. Tidwell, and Terrell A. Barry, Pest Management Analysis and Planning Program.

State of California Environmental Protection Agency; Department of Pesticide Regulation
Division of Enforcement, Environmental Monitoring, and Data Management
Environmental Monitoring and Pest Management Branch
1020 N Street, Sacramento; California 95814-5624.

This report provides findings from a study conducted by CDPR, in cooperation with the California Department of Education, which examined pest management programs in California's public school districts. The report provides an overview of current practices and recommended improvements to these practices.

EPA–IPM IN SCHOOLS

http://www.epa.gov/pesticides/ipm/

This Web site provides access to EPA and State resources addressing IPM in schools.

PESTICIDE USE AT NEW YORK SCHOOLS: REDUCING THE RISK

http://www.oag.state.ny.us/press/reports/pesticide_school/pesticide_school.html

This document contains a review of the issue and recommendations. Appendix 3 of the document contains a listing of information about IPM in schools.

POISONED SCHOOLS: INVISIBLE THREATS, VISIBLE ACTIONS. MARCH 2001

This report is a combined effort involving member organizations of the Child Proofing Our Communities: Poisoned School Campaign. This is a locally based, nationally connected campaign to protect children from exposure to environmental health hazards in schools, homes, and communities. It is available on the internet at http://www.beyondpesticides.org, or can be ordered from:

Child Proofing Our Communities Campaign
c/o Center for Health, Environment and Justice
P.O. Box 6806; Falls Church, VA 22040
Telephone: 703-237-2249
E-mail: childproofing@chej.org

THE SCHOOLING OF STATE PESTICIDE LAWS–2002 UPDATE: A REVIEW OF STATE PESTICIDE LAWS REGARDING SCHOOLS.

Kagan Owens and Jay Feldman. *Pesticides and You*, 20:2:16–23, 2000.

Beyond Pesticides/National Coalition Against the Misuse of Pesticides. (This report updates two earlier reports released in *Pesticides and You*, 20;2:16-23, 2000 and 18:3:1998.) Available from Beyond Pesticides at http://www.beyondpesticides.org/

G.8.3 Vector Control and Pest Eradication Programs

The use of pesticides in the eradication of disease vectors and economic pests raises issues and concerns about potential public exposure. Several publications and resources are provided here. This is an area with many available sources of information, so this listing is just a starting point.

American Mosquito Control Association

http://www.mosquito.org/

This professional association provides links to educational materials, State and local affiliate organizations, and other relevant resources.

Animal and Plant Health Inspection Service

http://www.usda.gov/

Provides information about pest eradication programs.

Department of Defense Pest Management Homepage

http://www.afpmb.org/

This site contains access to directive and guidance documents on pest control from the Armed Forces Pest Management Board. The Contingency Pest Management Guidance document contains information that may be useful for PPSPs involved in the public health aspects of vector control.

Toxicological Profile for Malathion Draft for Public Comment

http://www.atsdr.cdc.gov/toxprofiles/tp154.html

U.S. Department of Health and Human Resources, Agency for Toxic Substances and Disease Registry, Atlanta, Georgia, September, 2001.

Emerging Infectious Disease Journal

http://www.cdc.gov/ncidod/EID/index.htm

This journal can be searched for articles on mosquito control and other relevant topics.

EPA and Mosquito Control

http://www.epa.gov/pesticides/factsheets/skeeters.htm

This site has fact sheets related to mosquito control (larvicides, repellents, malathion, naled, pyrethrins, etc.), a joint EPA/CDC statement on mosquito control, and links to other sites.

HUMAN HEALTH SURVEILLANCE DURING THE AERIAL SPRAYING FOR CONTROL OF THE AMERICAN GYPSY MOTH ON SOUTHERN VANCOUVER ISLAND, BRITISH COLUMBIA, 1999

http://www.caphealth.org/btk journals.html

A report to the Administrator, Pesticide Control Act, Ministry of Environment, Lands and Parks, Province of British Columbia. Prepared by the Capital Health Region, Office of the Medical Health Officer, Director of Research, December 31, 1999.

WEST NILE VIRUS HOME PAGE–CDC DIVISION OF VECTOR BORNE INFECTIOUS DISEASES

http://www.cdc.gov/ncidod/dvbid/westnile/index.htm

Provides much information about West Nile virus infections, including background information, entomology, vertebrate ecology, virology, and surveillance.

G.9 State PPSP Contact Information

The following list of State contacts is not a comprehensive list of all States that may collect some pesticide-related illness surveillance data. It includes those that routinely produce annual reports and/or who are partners with EPA and NIOSH in efforts to standardize and enhance pesticide poisoning surveillance.

ARIZONA
Arizona Department of Health Services
Office of Environmental Health
3815 North Black Canyon Highway
Phoenix, AZ 85028
Telephone: 602-230-5830
http://www.hs.state.az.us/phs/oeh/invsurv/pesticide/index.htm

CALIFORNIA
California Department of Health Services
Occupational Health Branch
1515 Clay Street Suite 1901
Oakland, CA 94612
Telephone: 510-620-5757   Fax: 510-620-5743
http://www.dhs.ca.gov/ohb/AgInjury/Default.htm

California Department of Pesticide Regulation
Pesticide Illness Surveillance Program
1020 North Street, Room 200
Sacramento, CA 95814
Telephone: 510-540-3547   Fax: 510-540-3472
http://www.cdpr.ca.gov
Program Brochure:
http://www.cdpr.ca.gov/docs/whs/pisp.htm and click on *About the program.*

FLORIDA
Florida Department of Health
Bureau of Environmental Epidemiology
4052 Bald Cypress Way Rm. 215L
Tallahassee, FL 32399-1712
Telephone: 850-245-4115   Fax: 850-922-8473
http://www.doh.state.fl.us/

IOWA
Iowa Department of Health
321 E. 12th Street Lucas State Office Bldg.
Des Moines, IA 50319-0075
Telephone: 515-281-6596   Fax: 414-242-6384
http://www.idph.state.ia.us/eh/toxicology_env_health.asp#pesticide

LOUSIANA
Louisiana Office of Public Health
Environmental Epidemiology and Toxicology
325 Loyola Avenue Suite 210
New Orleans, LA70112
Telephone: 504-568-8322   Fax: 504-568-5815
http://www.dhh.state.la.us

MICHIGAN
Michigan Department of Community Health
Division of Environmental and Occupational Epidemiology
3423 N. Martin Luther King, Jr. Blvd.
P.O. Box 30195
Lansing, MI 48909
Telephone: 517-335-8761
http://www.michigan.gov/mdch/

MISSISSIPPI
Mississippi State Department of Health
2423 North State Street
Felix J. Underwood Building
Jackson, MS 39216
Telephone: 601-960-7725

NEW MEXICO
New Mexico Department of Health Services
1190 St. Francis Ave
Santa Fe, NM 87502
Telephone: 505-476-3583   Fax: 505-476-3589
http://www.health.state.nm.us/Web site.nsf/frames?ReadForm

Border Health
1170 North Solano Drive Suite L
Las Cruces, NM 88001
Telehone: 505-528-5156   Fax: 505-528-6045

NEW YORK
New York Department of Health
Bureau of Occupational Health
547 River Street Flanigan Square - Rm 230
Troy, NY 12203
Telephone: 518-402-7900   Fax: 518-402-7909
http://www.health.state.ny.us/nysdoh/pest/pesticid.htm

OREGON
Environmental and Occupational Epidemiology
Department of Health Services
800 NE Oregon St, # 827
Portland, OR 97232 2162
Telephone: 503 731 4025   Fax: 503 872-5398
http://www.ohd.hr.state.or.us/pesticide/index.cfm

TEXAS
Texas Department of State Health Services
Environmental & Occupational Epidemiology Program
1100 West 49th Street
Austin, TX78756
Telephone: 512-458-7111   Fax: 512-458-7699
http://www.dshs.state.tx.us/epitox/pest.shtm

WASHINGTON
Washington State Department of Health
Office of Toxic Substances
7171 Clean Water Lane, Building 4
Olympia, WA98501
Telephone: 360-236-3361   Fax: 360-586-4499
http://www.doh.wa.gov/ehp/ts/pest.htm

G.10 Federal Agency Contact Information

G.10.1 U.S. Department of Agriculture (USDA)

The USDA programs relevant to pesticide illness surveillance were described in Chapter 5, "Case Intake and Investigation" and in Section G.5. The agency descriptions are not repeated here. USDA also maintains the National Agricultural Library, which is a national network that provides a wide variety of information about agricultural issues. Information about the USDA services described here are available at http://www.usda.gov/AgenciesandOffices/.

Animal Plant Health Inspection Service

For telephone contact information, look in your local phone directory, or search using the term *state plant regulatory officials* at http://www.aphis.usda.gov/ppq/searchpage.html.

Cooperative State Research Education and Extension System (CSREES)

Local offices of CSREES are available in the Federal or county government listings of your local telephone directory, usually under the heading of Agricultural Extension Service, Extension Service, or Farm Services. The main State contacts are at land grant universities. A listing of State universities with land grant university status is available at http://www.nasulgc.org/About_Nasulgc/about_nasulgc.htm.

Contact: CSREES; USDA; Washington, DC 20250-0900
Telephone: 202-702-3029
Fax: 202-690-0289

Federal Grain Inspection Service (FGIS)

http://www.usda.gov/gipsa/

Federal Grain Inspection Service; Grain Inspection and Stockyards, Administration; STOP 3601
1400 Independence Avenue; Washington, DC 20250-3601

For telephone contact information, look in your local phone directory or search using the term *The Animal Plant Health Inspection Service* at http://www.usda.gov/gipsa/.

National Agricultural Statistics Service (NASS)

NASS Hotline 1-800-727-9540
NASS Census Division can be reached by calling 1-800-523-3215
Listings for NASS State offices are available at http://www.usda.gov/nass/sso-rpts.htm.

G.10.2 U.S. Department of Education—Office of Migrant Education

This office supports educational programs for migratory children. The local offices can be partners in PPSP education and outreach efforts. This program has produced a *Directory of Services for Migrant and Seasonal Farmworkers and Their Families* (1999). The directory is now maintained as an online database that is available at http://www.ael.org/eric/mied/. For more information and access to State offices, see http://www.ed.gov/about/offices/list/oese/ome/index.html.

G.10.3 U.S. Department of Labor (DOL)

OSHA

http://www.osha.gov/

This agency's Web site provides access to regulations and interpretations, technical documents, and directories of Federal and State program offices.

Wage and Hour Division

This division of DOL has responsibilities for enforcement of the Fair Labor Standards Act and the National Migrant and Seasonal Agricultural Worker Protection Act (MSPA). Information about Federal rules on labor contractors, wage and hour issues, child labor, the HB2A program, and a listing of district offices are available at http://www.dol.gov/esa/whd/. To locate a local Wage and Hour Office or get additional information, call toll-free 1-866-487-9243.

G.10.4 U.S. Environmental Protection Agency (EPA)

EPA/Office of Pesticide Programs 7506 C; Ariel Rios Building
1200 Pennsylvania Ave, NW; Washington, DC 20460
Telephone: 703-305-7666
Fax: 703-308-2962

http://www.epa.gov/pesticides

Regional pesticide contacts are available at this site as well as access to the EPA FIFRA rules (FQPA, WPS).

To receive weekly e-mail updates from this office, you can subscribe to the mailing list by visiting http://www.epa.gov/oppfead1/cb/csb_page/form/form.html.

An EPA-OPP contact that is useful for PPSPs is provided below. Questions about general pesticide issues, exposure concerns, and reporting of clusters and unusual illnesses and injuries from pesticides should be directed to either of the two numbers below.

EPA/ Office of Pesticide Programs; Health Effects Division 7509C
401 M Street SW; Washington DC 20460
Telephone: 703-305-7576 or 703-305-5336
Fax: 703-305-5147

G.10.5 U.S. Department of Health and Human Services—Centers for Disease Control and Prevention (CDC)

National Center for Environmental Health (NCEH), CDC

The Environmental Hazards and Health Effects Section of NCEH conducts investigations that increase knowledge of the relationship between human health and the environment and uses this understanding to develop national public policy and programs to prevent disease. NCEH studies ways to prevent or control health problems associated with exposure to air pollution, nuclear radiation, lead, and other toxicants, as well as hazards resulting from natural and technologic disasters. NCEH is a resource for environmental pesticide case surveillance and disease outbreak investigations. For general information about NCEH, see http://www.cdc.gov/nceh/.

The contact office for issues related to surveillance of pesticide-related illness and injury is:

National Center for Environmental Health; Health Studies Branch
Centers for Disease Control and Prevention
4770 Buford Highway, F46, Atlanta, GA 30341
Telephone: 770-488-3406
Fax: 770-488-3450

National Institute for Occupational Safety and Health (NIOSH), CDC

http://www.cdc.gov/niosh/topics/pesticides/

NIOSH is the Federal agency responsible for conducting research on occupational disease and injury. NIOSH investigates potentially hazardous working conditions upon request, makes recommendations on preventing workplace disease and injury, and provides training to occupational safety and health professionals. NIOSH provides funding and technical support to the Sentinel Event Notification System for Occupational Risks (SENSOR)-Pesticides program and supported the development of the SPIDER software for State-based surveillance of pesticide-related illness and injury. NIOSH also supports State-based surveillance of other occupational diseases and injuries. For general information about NIOSH programs, see http://www.cdc.gov/niosh/homepage.html.

The NIOSH contact for PPSPs is:

Project Officer, SENSOR-Pesticides
NIOSH/ Division of Surveillance, Hazard Evaluations and Field Studies
4676 Columbia Parkway, RB17; Cincinnati, OH 45226
Telephone: 1-800-356-4674
Fax: 513-533-8573

G.10.6 U.S. Department of Health and Human Services—Health Resources and Services Administration (HRSA)

Migrant Health Program (MHP), Bureau of Primary Health Care (BPHC)

The Migrant Health Act of 1962 was added to the Public Health Services Act. The Migrant Health Program provides medical and support services to migrant and seasonal farmworkers and their families. The MHP provides grant funds to public agencies and nonprofit organizations in 42 States and Puerto Rico for the development and operation of medical clinics. More information about the MHP and links to related resources are available at http://bphc.hrsa.gov/migrant/.

For information contact:

Division of Community and Migrant Health; Bureau of Primary Health Care
4350 East-West Highway, 7th Floor; Bethesda, MD 20814
Telephone: 301-594-4303
Fax: 301-944997

G.10.7 U.S. Food and Drug Administration (FDA)

The FDA is responsible for pesticide residue monitoring in foods as mandated by FIFRA and the Federal Food Drug and Cosmetic Act. Residue monitoring reports and other technical documents are available at http://www.cfsan.fda.gov/~lrd/pestadd.html.

G.11 Agricultural Safety and Health (Other Than Pesticides)

G.11.1 EPA

EPA and OSHA have collaborated to produce materials on heat stress in agriculture. A training guide, posters, and a laminated card are available on this topic from EPA or through the Government Printing Office. The document titles and document numbers are listed here.

A Guide to Heat Stress in Agriculture, 1994, Document No. 055B000B00474B9

Controlling Heat Stress in Agriculture, 1996, Document No. 055B000B00557B5 (Available in English or Spanish, 8"x4" cards with key items on recognizing and preventing heat stress.)

Controlling Heat Stress Made Simple/Maneras Sencillas de Controlar la Fatiga Causada por el Calor. Document No. 055B000B00544B3. Two-sided poster, available in English or Spanish.

G.11.2 NIOSH

NIOSH has many projects related to agricultural safety and health. To obtain information about this topic, visit: http://www.cdc.gov/niosh/topics/agriculture/.

NIOSH AGRICULTURAL HEALTH AND SAFETY CENTERS

NIOSH has funded eight Agricultural Health and Safety Centers throughout the country, which involve clinicians and other health specialists in the area of pesticide related illness and injury. A listing of the NIOSH supported centers with links to the individual center Web sites is available at http://www.cdc.gov/niosh/agctrhom.html.

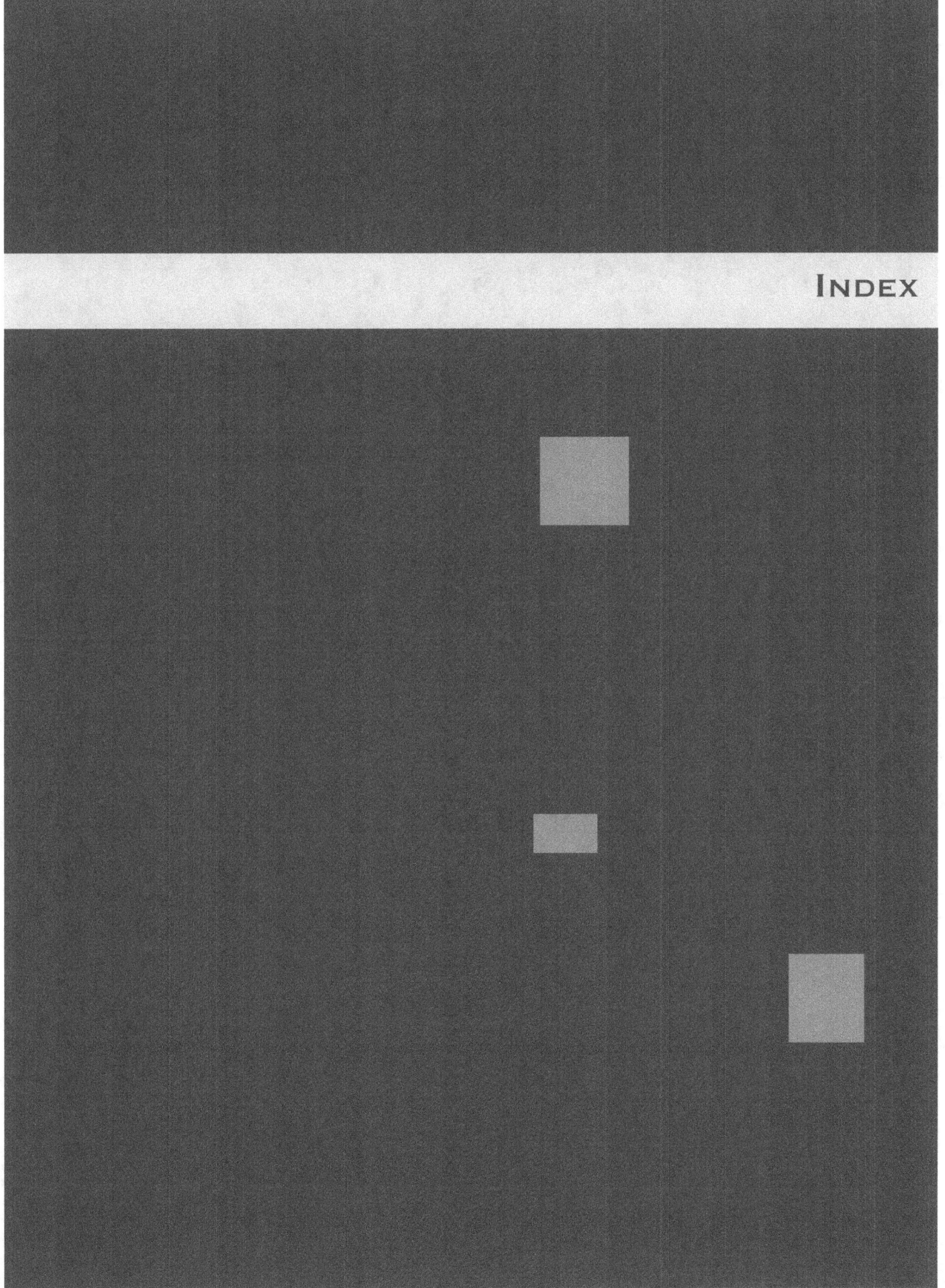

Index

INDEX

The following index is provided for readers of the *Pesticide-Related Illness and Injury Surveillance, A How-To Guide for State-Based Programs* to locate information by page number.

A

AAPCC, xi, 6, 13, 28, 90, 209, 211, 231, 237

active ingredients, 4, 6, 11, 55, 57, 76, 81, 183-184, 191-203, 217, 221, 238, 241

adjuvants, 5, 15, 20, 40, 54, 57

advisory committee, 54, 61, 93

advisory opinion, 219

affected person, 7, 14-17, 31-33, 39, 47-48, 50-52, 54-55, 58, 67, 71, 78, 87, 170, 174, 187

AFOP, xi, 245, 251

Agricultural Extension Service, 12

agricultural worker, 29, 31, 34, 82, 90, 106-107, 248-251, 258

ALT, xi, 194, 200

AMA, xi, 6, 97

analysis of data, 81-84, 87

ANR, xi, 247-248

ANSI /ASAE, xi, 89, 97

AOEC, xi, 235

APHIS, xi, 62-63

Arizona, 6, 111, 243, 251, 255

ASPCA, xi, 237

AST, xi, 194, 200

ATSDR, xi, 59, 62, 232-233, 235, 254

B

biological monitoring, 40, 89, 184

biological sampling, 70

biological specimen, 17, 20, 40, 51, 71, 184, 219

bivariate analyses, 81

BLS, xi, 23, 82

BOC, xi, 39

BPHC, xi, 249-250, 260

C

California, 6, 13-14, 17, 34, 63, 82, 89-91, 97-99, 105-107, 111, 119, 229, 233, 236, 239-241, 243, 246-249, 251-253, 255

case
    ascertainment, 27-34, 49
    classification, 3, 12, 40, 75-78, 183-205
    closure, 38, 47-48, 75-78, 119
    definition, iii, 14, 21, 27, 29,37, 40 47-48, 75-77, 83, 93, 183-205, 209, 211
    follow-up, 20, 21, 47-54, 116, 169
    intake, 38, 47-50,
    investigation, 3, 11-13, 31, 33, 37, 40, 47-64, 87-88
    is reportable, 184
    management, 20, 27, 231
    report, 3, 11, 14, 27, 32, 37-38, 50-51, 61, 77, 93, 170, 231
    series, 76, 81, 84, 92, 184, 186-187

CCOHS, xi, 230

CDC, xi, 7, 11, 13, 18-19, 38-39, 43, 90-93, 97-98, 100, 105-108, 119, 141, 187, 255, 259

CDPR, xi, 4, 13, 40, 90, 98, 229, 240-241, 243, 246, 252-253, 255

CFR, xi, 98

chemical class, 40, 81-82, 191

children, 4-5, 53, 223, 235, 248-249, 253, 258

CIRS, xi, 249

class of pesticide, 67, 183-184, 187, 191-192

clusters, 52-53, 55, 258

CME, xi, 31, 232-233

coding systems, 38-41
   (*see also* industry/occupation codes)
   hospital, 32-34
   industry, 39
   occupation, 14, 28, 31, 39
   variable, 38-41

confidential information, 20, 38, 41-42, 58, 68

confidentiality, 15, 17, 19, 29, 33, 38, 42, 51, 54-55, 57-58, 69, 84

CPS, xi, 23, 39, 82

CPSC, xi, 64

CSREES, xi, 62, 257

CSTE, xi, 6, 23, 75, 98

Current Population Survey (*see* CPS)

D

DA, xi, 17, 53, 57-59, 63, 90-92

data
   analysis, 37-38, 41, 81-84, 87
   collection, 3, 23, 37-38, 42, 48, 50, 54, 82, 92, 119, 141, 243
   dissemination, 38, 81, 84,
   entry, 3, 38, 42, 119, 141
   management, 12, 27, 37, 42, 119, 141
   optional variables, 40

DDT, xi, 4, 107

death, 5, 7, 23, 28, 61, 64, 67, 106, 172, 174, 179, 210

DEET, xi, 189, 195

demographic, 27, 29, 33, 50, 53, 229, 249-250

denominator data, 16, 21, 23, 82

DHHS, xi, 18-19, 237, 249

diagnosis, 11, 15-16, 20, 27-29, 32, 40, 76, 171, 183-184, 189, 234-235

discrimination, employment, 55

disinfectants, 4, 15, 31, 76, 176, 231

DO, xi, 187

DOL, xi, 11, 14, 59, 248, 250, 258

dose, 5, 186, 190-191, 240

DOT, xi, 64

E

education, 3, 31, 37, 60, 62-63, 81, 87-88, 230-233, 237, 245, 247, 251-253, 257-258

emergency department, 32, 105

emergency response, 68, 187, 238

employee representatives, 174

employment data
   rate-based surveillance with, 82
   sources of, 39

employment discrimination, 55

EMT, xi, 76, 184, 187

engineering control, 78, 87-89, 91

environment, 4-5, 51, 56, 59-61, 70-71, 88, 90, 92, 217-219, 221, 224, 239, 253, 259

EPA, xi, 4, 6-7, 13, 32, 34, 40, 43, 48, 50, 52, 55-59, 61-62, 70, 76, 88-91, 97-100, 170-171, 187, 203, 211, 217-218, 220-224, 230-232, 235, 237-241, 243, 245-246, 251-255, 258, 260

epidemiology, 31, 234

EU, xi, 240-241

EUP, xi, 220

evaluation, 31, 87, 92-93

exposure

   acute, 5, 61, 75-76, 83, 185, 190-191, 236

   chronic, 3, 5

   intentional, 39

   mixed, 3, 57

   nonoccupational, 27, 50-51, 58-59 61, 107, 183-188, 230, 236, 252-255

   occupational, 5, 14, 16-17, 21, 23, 28-29, 31, 51-52,54, 58-59, 69, 76 105-107, 183-188, 236

   unintentional, 5, 61, 106

exposure analyses, 70

extension service, 62, 230-231, 247, 252, 257

EXTOXNET 62, 231, 239, 240, 252

F

FAA, xi, 63

FAQ, xi, 187-191

farmworker, 28, 51-54, 82, 97, 99, 106, 119, 217, 229, 244-246, 248-252, 258, 260

FBI, xi, 64

FDA, xi, 64, 223

FFDCA, xi, 223

FGIS, xi, 62, 63, 257

field investigation, 19-20, 22, 67-71, 82, 87, 92, 161-164, 173-175115, 161, 173

FIFRA, xi, 52, 56, 58-59, 67, 70, 76, 183, 211, 217-225, 238, 258, 260

FJF, x, 249, 251

Florida, 6, 14, 28, 53, 63, 90-91, 236, 249, 251, 255,

follow-up, 47-53, 59, 75, 78, 87-88, 116-118, 161-164

forms, 3, 33, 37, 58, 115-164

FQPA, xi, 223-224, 258

fumigants, 59, 63, 105-106, 177, 197

functional class, 40, 82, 191

funding, 7, 13, 27, 43, 67, 70, 84, 87, 259

G

GAO, iii, xi, 6, 13, 99, 249

gender (*see* Sex)

GIS, xi, 83, 84

H

HCP, xi, 11, 14, 16, 27-33, 40-41, 47-48, 51, 53-55, 57, 61, 63, 77, 81, 84, 87, 141, 170- 172, 185-189, 191, 210, 230-237, 252

HDD, xi, 33

health department, 48, 51, 54-55, 57, 59

HIPAA, xi, 18-19, 172

HRSA, xi, 260

HSDB, xii, 239

HSEES, xii, 59

I

IARC, xii, 239-241

IATP, xii, 246

ICD, xii, 29, 32-33

ICD10, 29, 32-33

ICD9, 29, 32-33

ICU, xii, 23

identifiers, 18-19, 21, 33, 40, 43, 171-172, 174

illegal, 62, 64, 106

ILO, xii, 239

incidence, 11, 20, 23, 52, 176, 179

indicator, 4, 23

industry and occupation codes, 39

inert ingredients, 15, 55-57, 238-239

ingestion, 15, 27, 67, 105-106

injury, pesticide, xiii, 3-7, 18, 29, 31, 75, 183, 185, 188,

insurance data (*see* workers' compensation data)

interagency agreement, 29, 60-61, 71

interagency cooperation, 19, 218

interagency coordination, 48, 60-61, 225

intervention, iii, 11, 16, 47-48, 60-62, 87-93, 172, 209, 249

interviews, 12, 47-48, 50-55, 67, 69, 75, 93, 119, 141, 173-175

investigation, case 11, 16-21, 31-33, 37-38, 47-58, 60-64, 87-88, 225

investigation, field, (*see* field investigation)

IPCS xii, 239

IPM, xii, 12, 62, 78, 88-89, 93, 230, 247, 252-253

IRIS xii, 240-241

L

label, 5-7, 40, 55-56, 58, 61, 70-71, 90, 92, 217-224, 239, 241-242

laboratories, 16, 18, 20, 33-34, 47, 71

LAN, xii, 41

latency, 190

LDH, xii, 194, 200

letters, 54, 169-175

local health department, 48, 51, 55

lost time, 29, 40, 209-210, 214

Louisiana, 6, 17, 53, 108, 249, 256

LSC, xii, 251

M

MD, xii, 187

management
confidential information, 20, 42, 54-55, 58
data, 12, 27, 31, 37, 41-43 119, 141
integrated pest management, (*see* IPM)
manufacturer, 6, 48, 57, 61, 70, 76, 91-92, 217, 230-231, 237, 241-242

mapping of data, 83-84

maps, 63, 83-84, 244

MCN, xii, 250

medical record, 16, 48, 53-54, 67

MHP, xii, 260

Michigan, 6, 111, 239, 256

Migrant farmworker, 14, 52, 54, 248-252, 258, 260

Migrant Clinicians Network, 250

Migrant Health Program, 249, 260

migrant worker, (*see* migrant farmworker)

MMWR, xii, 13, 81, 105-108

MSDS xii, 55-57, 241-242

MSPA, xii, 248, 258

N

NAICS, xii, 39

NAIN, xii, 231

NAS, xii, 100

NASS, xii, 244, 257

NAWS, xii, 14, 250

NCEH, xii, 7, 13, 43, 61-62, 187, 259

NCFH, xii, 250, 252

NCHS, xii, 39, 100

NEETF, xii, 237

New Jersey, 111, 243, 251

New Mexico, 111, 256

New York, 6, 14, 20, 41, 63, 82, 100, 105-106, 112, 224, 243, 245, 251, 253, 256

NIH, xii, 240-241

NIOSH, iii, xii, 7, 13, 23, 38-41, 43, 48, 61-62, 75, 90, 93, 99-100, 119, 141, 174, 187, 209, 211, 236, 255, 259-261

nonoccupational exposure, (*see* exposure, nonoccupational)

notifiable conditions, 29, 33

notification, 48, 53, 55, 61-62, 76-78, 183, 246, 259
    local health department, 48, 55
    to employer, 77-78
    to EPA, 48, 61-62
    to manufacturer, 77-78
    to NCEH, 61-62, 76-77
    to NIOSH, 48, 61-62, 76-77
    to physician, 28-29, 31, 34, 76-77

NPHSS, xii, 40, 75, 77, 98

NPIC, xii, 12, 13, 57, 90, 187, 231, 232, 242, 252

NPMMP, xii, 12, 232

NSCEP, xii, 239, 243

NTP, xii, 240-241

NTSB, xii, 165

NUBC, xii, 33

NYDEC, xii, 243

O

objectives, 11, 37, 92

occupational exposures, (*see* exposure, occupational)

ODA, xii, 59, 92 245

OEHHA, xii, 233

OHSU, xii, 245

OPP, xii, 187, 240-241, 252, 258

Oregon, 6, 14, 19, 59-61, 87-88, 91-92, 99-101, 112, 231, 239, 243, 245, 251, 256

OSH Act, xi, 55, 67, 224

OSHA, xii, 32, 59, 64, 69-70, 100, 173-174, 224, 236, 258, 260

OSU, xii, 232, 245

outreach, 3, 31, 61, 63, 81, 83, 88, 93, 251, 258

P

PA, xii, 187

PANNA, xii, 55, 236, 240-241, 243

PAPR, xii, 70

PARC, xiii, 61, 88, 92, 100

passive surveillance, 20

patient identifier, 18-19, 21, 33, 40, 43, 171-172, 174

PCC, xiii, 12, 14, 17-18, 20, 23, 27-28, 47, 57, 61, 176-180, 231, 239 person, place, and time, 81

personal identifiers, 18-19, 21, 38, 40, 43, 171-172, 174

pesticide application, 5, 17, 83-84, 234, 236, 243-244, 247

pesticide applicator, 15-16, 28, 37, 41, 48, 53-55, 57-58, 62-63,165, 185, 188-189, 234, 243, 245-246

pesticide manufacturers, (*see* manufacturer)

pesticide use, 4-7,13, 15, 16, 19

pesticide use reporting, 16, 243-245

PHI, xiii, 18, 19

PIRT, xiii, 61, 112

PISP, xiii, 90

POISINDEX®, 55

Poison Control Centers, (*see* PCC)

policy, 17, 33, 56, 83, 259

population 23, 82-83, 87, 92 176, 179-180

PPE, xiii, 6, 39, 68, 69-70, 78, 90, 161-164, 222

PPIS, xiii, 40, 55, 240

PPSP, xiii, 6-7, 11-14, 16-17, 27-28, 32-33, 37- 38, 40-43, 47-48, 50-51, 53-64, 67-68, 70-71, 75, 77, 81-84, 87-88, 90-93, 116, 161, 169, 229, 231, 237, 239-242, 248, 250, 254-255, 258-259

predictive value positive, 92

prevalence, 6, 83

protocol, 20, 42, 48, 52, 55, 57-59, 67, 70, 75, 87, 92, 189
   case triage, 42
   data system backup, 42
   investigation, 87

PVC, xiii, 70

Q

questionnaire, 50, 52, 54, 115, 119-141, 250

R

RN, xiii, 187

rate(s), 11, 16, 21, 23, 39, 42, 63, 82-83, 176, 179-180, 204-205, 244

RCW, xiii, 17, 20,, 112

records, 16, 32, 42, 47, 48, 53, 54, 67, 69, 83, 115-116, 169, 172, 174, 186, 221, 234
   duplicate, 42, 179
   electronic, 42
   emergency department, 32
   hard copy, 42
   medical, 16, 48, 53, 54, 67, 169, 172, 174, 187, 189
   pesticide application, 83, 186, 234, 243
   tracking, 38, 40-41, 43, 50, 53, 58, 61, 77, 82, 93, 115-116, 173, 211

REDs, xiii, 238

referral, 15, 17, 20, 32-34, 41, 47-48, 51, 57-59, 62, 68, 84, 116, 171, 231-232

reportable condition, 17, 19, 31, 33, 171

reporting laws, 13, 17-20, 22

reporting rule, 13 -17, 20-21, 27, 29, 33-34, 67, 111-112

reporting statutes, 13-14, 16-17, 19-20, 27, 29,109, 111

reports 3, 5-6, 13-14, 17-18, 20-21, 27-29, 31-32, 34, 37, 41-42, 47-55, 59-63, 67, 75-78, 81, 83-84, 91, 93, 107, 109, 115-116, 161-168, 173-174, 181-185, 188-191, 221, 223, 229, 231, 243-244, 246, 248-253, 255, 260
   annual, 13, 61, 81, 83-84, 231, 243-244, 255

case, 3, 5-6, 14, 17-18, 20-21, 27-29, 31-32, 34, 37, 42, 47-55, 60-62, 97, 75-77, 91, 93, 115-116, 173-174, 184-8, 188, 222, 225, 231
 data, 41-42, 81, 83-84, 93, 161-169, 243-244, 249-251
 GAO, 6, 13-14, 251
 investigation, 75, 78, 93

representativeness, 92

risk factors, iii, 5-6, 81, 82, 83, 87, 89, 93

route of exposure 5, 27, 39

S

samples, 20-22, 40, 53, 55, 70-71, 185, 191

SAS, xiii, 41

schools, 16, 19, 28, 81, 83, 88, 229, 252-253

SDWA, xiii, 224

sensitivity, 14, 29, 52, 92

SENSOR, xiii, 3, 39, 83-84, 97, 119, 141, 259

severity, 3, 23, 27, 40, 51, 57-58, 77, 81, 140, 160, 165, 175, 207, 209-211

sex, 5, 16, 23, 38, 81-82, 179

signs, 5, 23, 28, 40, 76, 171, 134-136, 140, 156-158, 160, 183-184, 186-187, 189-190, 192-203, 209-210, 212-214, 236

site evaluation, 48, 51, 69

site inspection, 48, 54, 57, 65, 67-71, 75, 78, 93, 173

site investigation, 47, 82, 87, 173-175

SOC, xiii, 39

SPIDER, xiii, 3, 38, 40-43, 55, 83-84, 100, 119, 141, 241, 259

SPPC, xiii, 119

staff, 12-13, 32, 37-38, 42, 47-48, 50-55, 59, 67-71, 88, 92, 161, 171, 218, 230-231, 234, 245-246, 248, 250

staffing, 11-13

standardized variables, 3, 37-42, 209

State health departments, 3, 17, 19, 32

suicide, 4, 28, 76, 184

support, iii, 13, 16-17, 19, 27, 50, 81, 232, 251, 259-260

surveillance program, iii, ix, 3, 6-7, 11-14, 16-17, 21, 27-29, 32, 37, 40, 47-48, 67, 75, 77, 84, 87, 90-93, 119, 141, 170, 171, 217-218, 225, 255-256

surveillance system, 3, 4, 6-8, 11, 16-17, 31-33, 37, 40, 47-48, 50, 75-77, 90, 92-93, 181, 183-184
 flow diagram, 48

symptoms, 5, 23, 28-29, 31, 40, 50, 52-53, 76-77, 134-136, 140, 156-158, 160, 183-184, 186-187, 189-190, 192-203, 209-210, 212-213, 234-237

T

TBTO, xiii, 91-92

TESS, xiii, 4, 6, 28, 90

Texas, xiv, 6, 14, 20, 22, 60, 106, 112, 249, 256

timeliness, 6, 17, 23, 27, 29, 31, 33, 41, 42, 47, 61, 81, 84, 92

toxicology, iii, 12-13, 62, 64, 76-77, 140, 160, 184, 186-187, 229-231, 233, 237-241, 252

tracking, 15, 33, 38, 40-41, 43, 50, 53, 58, 61, 77, 82, 93, 115-116, 173, 184, 222

training, 12, 31, 38-39, 42, 50-52, 54-55, 59, 62, 68, 69, 71, 78, 91, 218, 223, 230, 233, 236-237, 245-247, 249-251, 259-260

trends, 11, 61, 81, 83, 107, 244
   time, 83

U

UB-92, xiii, 33
underreporting, 14
UNEP, xiii, 239
USDA, xiii, 62-63, 222, 223, 234, 243-244, 257
use data, (*see* pesticide use reporting)

V

variables, 3, 29, 37-42, 81, 119, 141, 209
   standard variables to be collected by PPSPs, 38

W

Washington State, xiii, 6, 13-14, 17, 19-20, 34, 53, 59-62, 67 83, 88-92, 106,112, 249-251, 256
WHO, xiii, 231, 233, 239-241, 251, 255
WHS, xiii, 246,
worker representatives, 68, 78, 173
workers' compensation, 29, 31, 33, 248-249
workers' compensation data 13-14, 17-18, 29, 30, 31, 33, 47
workers' compensation records, 47
worksite inspections, 67-71, 78, 173-175
WPS, xiii, 40, 58-59, 89, 222-223, 258
WSDA, xii, 91-92
WSDOH, xiii, 91, 92

www.ingramcontent.com/pod-product-compliance
Lightning Source LLC
Chambersburg PA
CBHW081720170526
45167CB00009B/3639